NUREG-1850

FREQUENTLY ASKED QUESTIONS ON LICENSE RENEWAL OF NUCLEAR POWER REACTORS

Final Report

Manuscript Completed: February 2006
Date Published: March 2006

Division of License Renewal
Office of Nuclear Reactor Regulation
U.S. Nuclear Regulatory Commission
Washington, DC 20555-0001

Abstract

With the increase in the number of power reactors submitting applications for renewed licenses, the U.S. Nuclear Regulatory Commission (NRC) staff realized that members of the public near different power reactor sites had similar questions related to the license renewal process and the risks associated with renewing the licenses. This report, through a question-and-answer format, provides staff responses to frequently asked questions on the license renewal process for commercial, nuclear power reactors. The questions were taken from a variety of sources over the past several years, including written inquiries to the NRC and questions asked at public meetings and during informal discussions with the NRC staff. In responding to the questions, the NRC staff attempted to provide the answers in a clear and non-technical form.

This document contains a definition of license renewal including information related to the timing and scheduling of the license renewal process. It also discusses the NRC's role in reviewing, approving, or denying license renewal and the regulatory basis for the review. Because the public usually encounters the license renewal process in conjunction with the environmental review, this document is primarily focused on the environmental review process and on related issues such as alternatives to license renewal and human health issues. However, other aspects of license renewal are addressed, including questions related to the safety reviews, the storage and disposal of spent nuclear fuel, and security issues. There are also responses to questions related to public involvement and to finding sources of additional information on the license renewal process.

Contents

Tables

Figures

Abbreviations

ACRS	Advisory Committee on Reactor Safeguards
ADAMS	Agency-wide Documents Access and Management System
ASLB	Atomic Safety Licensing Board
ASME	American Society of Mechanical Engineers
CEQ	President's Council on Environmental Quality
CLB	current licensing basis
CFR	Code of Federal Regulations
DOE	U.S. Department of Energy
EA	environmental assessment
EIS	environmental impact statement
EDO	Executive Director for Operations
EPA	U.S. Environmental Protection Agency
FACA	Federal Advisory Committee Act
FAQ	frequently asked questions
FSAR	final safety analysis report
GALL	Generic Aging Lessons Learned
GEIS	*Generic Environmental Impact Statement for License Renewal of Nuclear Plants*
GPO	Government Printing Office
ICRP	International Commission on Radiological Protection
IPs	Inspection Procedures
IPE	individual plant evaluation
IPEEE	individual plant evaluation of external events
ISFSI	independent spent fuel storage installation
LLTF	Lessons Learned Task Force
MAB	maximum attainable benefit
MC	Manual Chapter
NCI	National Cancer Institute
NCRP	National Council on Radiation Protection and Measurements
NEPA	National Environmental Policy Act
NPDES	National Pollutant Discharge Elimination System
NRC	U.S. Nuclear Regulatory Commission
NRR	Nuclear Reactor Regulation
ODCM	Offsite Dose Calculation Manual

PARS	publicly available records system
PFS	Private Fuel Storage, LLC
PSA	probabilistic safety analysis
SAMA	severe accident mitigation alternative
SEIS	supplemental environmental impact statement
SER	safety evaluation report
SHPO	State Historic Preservation Officer
SI	International system of units
SRP	standard review plan
SSCs	systems, structures and components
Sv	sieverts
TIs	Temporary Instructions
UNSCEAR	United Nations Scientific Committee on the Effects of Atomic Radiation
U.S.	United States

Introduction

The Atomic Energy Act of 1954 (as amended) allows the U.S. Nuclear Regulatory Commission (NRC) to issue licenses for commercial power reactors to operate for up to 40 years. The NRC regulations allow for the renewal of these licenses for up to an additional 20 years beyond the initial licensing period depending on the outcome of an assessment to determine whether the reactor can continue to operate safely and whether the protection of the environment can be ensured during the 20-year period of extended operation. The license renewal process includes reviewing a license renewal application, conducting the assessment, and then renewing the license. The NRC's review of a license renewal application proceeds along two tracks: one for safety issues and another for environmental issues. The license renewal process is defined by a clear set of regulations that are designed to ensure safe operation and protection of the environment during the period of extended operation.

The final rule containing the NRC regulations for the license renewal safety review was published in 1995 in Part 54 of the *Code of Federal Regulations* (10 CFR Part 54). The final rule revising the NRC's regulations in 10 CFR Part 51 to define the environmental issues for license renewal was published in 1996 and amended in 1999. In April 1998, Baltimore Gas and Electric became the first licensee to apply for license renewal for its Calvert Cliffs facility on Chesapeake Bay. Duke Energy Corporation followed suit in July 1998 when it sought renewed licenses for its Oconee nuclear units in South Carolina. As of the beginning of 2006, licenses for 22 facilities (for a total of 39 nuclear power reactor units) have been renewed.

As part of the environmental review process, the NRC conducts public meetings and solicits public comments and questions regarding the environmental portion of the license renewal application. During this process (and specifically during the public meetings) members of the public can ask questions related to license renewal. This document is a compilation of many of the pertinent questions that have been voiced since the first license renewal application was received. In responding to the questions, the NRC staff has attempted to provide the answers in a clear, non-technical form.

Section 1 of this document contains the responses to general questions related to the reason for renewing licenses, the timing and scheduling of the license renewal review, and the NRC's role in reviewing and either granting or denying the licensee's request to renew a license. Section 2 describes the regulatory basis for the license renewal review process. Section 3 discusses the NRC's safety review including the basis for the safety review and the NRC's safety inspection process. Section 4 contains questions and responses related to the environmental review, including the scope of the review, severe accident mitigation alternatives, storage and disposal of spent fuel, human health issues, and the identification of alternatives to license renewal. Section 5, the final section, discusses public involvement in the license renewal process. The final section also contains a bibliography of published materials related to license renewal.

Frequently Asked Questions for License Renewal

1.0 General Questions about License Renewal

This section contains general questions and responses related to license renewal. The questions are grouped into subgroups that include questions related to:

- the purpose of license renewal, including the benefits of renewal, the alternatives to renewal and a brief history of how the process of renewing licenses evolved,

- the NRC's and other agencies' role in reviewing and either granting or denying an applicant's request to renew a license, and

- the timing and scheduling of the license renewal application and review.

1.1 Purpose of license renewal

1.1.1 What is license renewal for power reactors?

License renewal is the process of conducting an assessment of the extended period of operation and renewing the license.

The Atomic Energy Act of 1954 (as amended) allows the NRC to issue licenses for commercial power reactors to operate for up to 40 years. This license is based on adherence of the licensee and facility to the appropriate regulations described in the *Code of Federal Regulations* (CFR), Title 10 (see response to Questions 2.3 and 5.2.6 for a further description of the CFR and how to obtain a copy). The NRC regulations allow the renewal of these licenses for up to an additional 20 years depending on the outcome of an assessment to determine whether the nuclear facility can continue to operate safely and whether the protection of the environment can be assured during the 20-year period of extended operation. There are no specific limitations in the Atomic Energy Act or the NRC's regulations restricting the number of times a license may be renewed. The process of conducting the assessment and renewing the license is termed "license renewal." The license renewal process includes a clear set of requirements, which are designed to assure safe facility operation and protection of the environment for up to an additional 20 years.

1.1.2 Why do nuclear power reactor licenses need to be renewed?

The original licenses for commercial nuclear power facilities were granted for a 40-year period, which was set by the Atomic Energy Act of 1954 and the NRC's regulations. It was imposed for economic and antitrust reasons rather than technical limitations of the nuclear facility.

Brunswick, Units 1 and 2

Studies and experience to date have shown that commercial nuclear power facilities can be safely operated for more than 40 years.

Studies and experience to date have shown that commercial nuclear power facilities can be safely operated for more than 40 years. As a result, the NRC has provided an option in Title 10 of the Code of Federal Regulations, which allows owners of nuclear power reactors to seek license renewal for up to an additional 20 years with no limitations on the number of times the license may be renewed. The decision whether to seek license renewal rests entirely with nuclear power reactor owners, and typically is based on the plant's economic viability and whether it can continue to meet NRC safety and environmental requirements. The NRC bases its decision on whether or not to renew a license on whether the facility will continue to meet the requirements for safe operation and whether the protection of the environment can be assured.

1.1.3 What are the benefits of license renewal to the facility licensee?

Once the license has been renewed, the licensee is then allowed to continue operating the facility for up to an additional 20 years (with no limitations on the number of times the license may be renewed) provided that it operates the facility safely and without unacceptable environmental impact. This is a benefit to the licensee because it provides the opportunity for continued energy production at an already existing facility, rather than requiring the construction of a new energy source with concurrent decommissioning of the existing facility.

1.1.4 What are the alternatives to license renewal?

The only alternative to renewing the license would be not to renew it. Not renewing the license would mean that the nuclear power facility would cease operation and would begin decommissioning at the end

of the original 40-year licensing period. The power lost by not renewing the operating license would need to be replaced by other sources of power, by demand-side management and energy conservation, by power purchased from other electricity providers or through some combination of these options. Further discussion of alternatives is given in Section 4.7 of this document.

1.1.5 Why are nuclear power reactors originally licensed for 40 years?

The Atomic Energy Act of 1954 originally specified that licenses for commercial power reactors be granted for a period not exceeding 40 years: "Each such license shall be issued for a specified period, as determined by the Commission, depending on the type of activity to be licensed, but not exceeding forty years, and may be renewed upon the expiration of such period." The 40-year licensing period was based on economic and antitrust considerations rather than on the technical limitations of the nuclear facility.

1.1.6 How did the process of renewing licenses evolve?

The provision in the Atomic Energy Act that stipulated a limit of 40 years for a commercial power reactor operating license also permitted renewal of the licenses.

In the early 1980's, the NRC staff recognized that it needed to identify the information required and the process to be used to determine whether to grant an extension to the operating license. In 1982, based on a widely attended workshop on nuclear power facility aging, the NRC established a comprehensive program for Nuclear Plant Aging Research. Based on the results of that research, a technical review group concluded that many aging phenomena are readily manageable and do not pose technical issues that would prevent nuclear power facilities from operating safely and efficiently for more than 40 years. In 1985, the NRC approved regulations that cover the safety and technical requirements for license renewal. These regulations (10 CFR Part 54) were adopted by the NRC and published in December 1991.

After considering ways to evaluate the environmental consequences of license renewal, the NRC chose to develop the *Generic Environmental Impact Statement for License Renewal of Nuclear Plants*, NUREG-1437 (GEIS), which covered impacts that were common to all or most nuclear power facilities. The GEIS that was published in 1996 allows the applicant and NRC to focus on those important environmental issues specific to each site being relicensed.

By the start of 2006, 22 facilities had their licenses renewed.

After further analysis, the development of a draft regulatory guide, a draft standard review plan for license renewal, and input during public workshops, the NRC amended the regulations published in 1991 to ensure a predictable and stable regulatory process that clearly defined the Commission's expectations for license renewal. In 1995, the amended rule was published. In 1996, the NRC published the final rule that revised 10 CFR Part 51, which contained the regulations for the environmental analysis related to relicensing.[1]

In April 1998, Baltimore Gas and Electric became the first licensee to apply for license renewal for its Calvert Cliffs facility on Chesapeake Bay. Duke Energy followed suit in July 1998 when it sought renewed licenses for its Oconee nuclear units in South Carolina. As of the beginning of 2006, licenses for 22 facilities (for a total of 39 nuclear power reactor units) have been renewed (see Table 1.1).

Calvert Cliffs, Unit 1 and 2

---

[1] The regulations in 10 CFR Part 51 were amended three times after 1996. The first amendment sought to change three typographical errors (November 3, 1997 published in the *Federal Register* at 62 FR 59276). The second amendment (September 3, 1999, published in the *Federal Register* at 64 FR 48507) reflected a new analysis of transportation issues that expanded the generic findings about the environmental impacts due to transportation of fuel and waste to and from a single nuclear power plant. This change revised the categorization of transportation of high level waste from Category 2 (site-specific) to Category 1 (generic). The third amendment (July 30, 2001, published in the *Federal Register* at 66 FR 39278) corrected a word that was inadvertently omitted from Table B-1 that made it appear that high level waste and spent fuel disposal were included in the environmental dose commitment to the U.S. population from the fuel cycle. However, the NRC's intent was for it to be excluded from the calculation since there was a separate finding for offsite radiological impacts (spent fuel and high level waste disposal) elsewhere in the table.

Table 1.1. Nuclear Power Reactor Facilities that Have Renewed Licenses or are Under Review

Facility	Date renewed license was issued
Calvert Cliffs, Units 1 and 2	March 23, 2000
Oconee Nuclear Station, Units 1, 2 and 3	May 23, 2000
Arkansas Nuclear One, Unit 1	June 20, 2001
Edwin I. Hatch Nuclear Plant, Units 1 and 2	January 15, 2002
Turkey Point Nuclear Plant, Units 3 and 4	June 6, 2002
North Anna Power Station, Units 1 and 2	March 20, 2003
Surry Power Station, Units 1 and 2	March 20, 2003
Peach Bottom Nuclear Reactor, Units 2 and 3	May 7, 2003
St. Lucie, Units 1 and 2	October 2, 2003
Fort Calhoun Station, Unit 1	November 4, 2003
McGuire Nuclear Station, Units 1 and 2	December 5, 2003
Catawba Nuclear Station, Units 1 and 2	December 5, 2003
H.B. Robinson Nuclear Plant, Unit 2	April 19, 2004
R.E. Ginna Nuclear Power Plant, Unit 1	May 19, 2004
V.C. Summer Nuclear Station, Unit 1	April 23, 2004
Dresden Nuclear Power Station, Units 2 and 3	October 28, 2004
Quad Cities Nuclear Power Station, Units 1 and 2	October 28, 2004
Farley, Units 1 and 2	May 12, 2005
Arkansas Nuclear One, Unit 2	June 30, 2005
D.C. Cook, Units 1 and 2	August 30, 2005
Millstone, Units 2 and 3	November 28, 2005
Point Beach, Units 1 and 2	December 22, 2005

Facility	Date application was received
Browns Ferry, Units 1, 2, and 3	January 6, 2004
Nine Mile Point, Units 1 and 2	May 27, 2004
Brunswick, Units 1 and 2	October 20, 2004
Monticello	March 24, 2005
Palisades	March 31, 2005
Oyster Creek Nuclear Generating Station	July 22, 2005
Pilgrim 1	January 27, 2006
Vermont Yankee	January 27, 2006

1.2 NRC's role in reviewing, approving or denying license renewal

1.2.1 Who makes the decision to renew a license for a nuclear power reactor?

The decision to seek a license renewal rests entirely with nuclear power facility owners.

It is helpful to distinguish "seeking a license renewal" from "granting or denying a license renewal." The decision to seek a license renewal rests entirely with nuclear power facility owners and typically is based on the facility's economic viability and the investment necessary to continue to meet NRC safety and environmental requirements. The NRC makes the decision to grant or deny a license renewal, based on whether the applicant has demonstrated that the environmental and safety requirements in the NRC's regulations can be met during the period of extended operation. If the applicant meets the requirements given in the regulations, then the NRC can be expected to approve renewal of the license.

1.2.2 Who makes the final decision to either approve or deny the request to renew the license?

The NRC Commission has delegated this authority to the Director of the NRC's Office of Nuclear Reactor Regulation (NRR).

The NRC makes the decision to grant or deny a license renewal, based on whether the applicant has demonstrated that the environmental and safety requirements in the NRC's regulations can be met during the period of extended operation.

1.2.3 Can state, county, local, Tribal, or other Federal (not NRC) agencies reject or disapprove of the license?

Because the licensing authority is the NRC's, only the NRC can approve or deny the application. However, the NRC will consider any comments provided by the state, county, local, Tribal or other Federal agencies during the period of the NRC's review and analysis. In addition, some of these agencies (such as state agencies) are in a position to specify conditions or reject permits that are required by the applicant, such as the National Pollutant Discharge Elimination System (NPDES) permit, which is usually administered by a state.

1.2.4 Do other Federal agencies review the license application? If so, which agencies?

The environmental scoping process invites other governmental agencies to assess whether or not they should be considered cooperating agencies under the regulatory structure afforded by the President's Council on Environmental Quality (CEQ). It also invites them to identify whether or not they have a particular expertise on an issue that may be invaluable to the NRC, or have consultation roles under other statutes that may have a bearing on site-specific issues.

A notice of the receipt of the license renewal application is posted in the *Federal Register* shortly after it is received by the NRC. The notice indicates where copies are available and how they can be obtained. Other Federal agencies that are interested in reviewing the application can obtain a copy and provide comments to the NRC during the scoping process or after publication of the draft site-specific supplement to the generic environmental impact statement (SEIS). However, the NRC is the lead agency, and as such

During the period of the NRC's review and analysis, the NRC considers comments provided by the state, county, local, Tribal or other Federal agencies.

has responsibility to review the license renewal application and develop the draft and final environmental impact statement.

1.2.5 Are Tribal nations provided with copies of the license renewal application for their review?

The NRC staff sends a letter directly to the leaders of the Tribal nations that may have an interest in the land occupied by or in the area surrounding the nuclear power facility. The letter informs the Tribal nations of the receipt of the license renewal application and provides them with information on how to obtain a copy. Instructions for providing comments to the NRC are also given in the letter.

1.2.6 Are non-governmental agencies provided with copies of the license renewal application for their review?

Any person or organization can obtain access to the license renewal applications (see response to Question 5.2.7). A notice of the receipt of the license renewal application is posted in the *Federal Register* shortly after the application is received by the NRC. The notice indicates where copies are available and how they can be obtained. A hard copy of the application is also made available for review at a public library close to the site. Also, electronic copies are accessible and downloadable from the NRC website.

1.2.7 Can the NRC say "no" to renewing a license?

Yes. As described in the regulations, based on the findings of its review, the NRC can deny an application to renew a license.

1.2.8 Has the NRC ever denied an application to renew a license?

The NRC can deny an application to renew a license based on the findings of its review.

To date, the NRC has approved all of the applications for license renewal for which the review has been completed. The NRC has found an application insufficient to start the review and has rejected an application. The NRC has also halted the review process until sufficient information is provided to continue the review. Although the NRC can deny a request to renew a license if the applicant did not provide appropriate or adequate information in its initial application, the NRC would identify the deficiencies and the applicant would be allowed to resubmit the application or provide additional information. This process can continue until the NRC concludes that the application is sufficient to complete the review.

The NRC has clearly defined the requirements for license renewal and the nuclear industry has the experience of many successful license renewal requests. Because of the cost and the commitment associated with an application, it is unlikely that an applicant would intentionally submit an application for license renewal that was so flawed that the NRC staff would issue a denial. Finally, if problems with systems, structures or components of the facility were identified during the review, the applicant would likely be able to make the required modifications or put in place a mitigation plan that would be acceptable to the NRC. Identified problems with structures, systems, or components would be addressed immediately, and any necessary changes made under the current operating license rather than waiting for the period of extended operation.

1.2.9 Does NRC approval of the application guarantee that the licensee will continue to operate the nuclear power facility?

Although a licensee must renew its license to operate a reactor beyond the term of the existing license, the possession of a renewed license is just one of a number of conditions that must be met to continue operation. Once a license is renewed, other factors and entities such as state regulatory agencies and the owners of the nuclear power facility will ultimately decide whether the facility will continue to operate. Economic considerations have a significant influence on the decision to continue operation.

From the perspective of the applicant and the state regulatory authority, the purpose of renewing a license is to maintain the availability of the nuclear facility to meet energy requirements beyond the current term of the facility's license.

1.2.10 Who makes the decision to actually continue operating the nuclear power facility during the license renewal period once the license renewal is granted?

It is possible that a license renewal application could satisfy the NRC's safety and environmental reviews and still not operate. This is because the NRC does not have a role in the energy-planning decisions of state regulators and licensee officials. From the perspective of the applicant and the state regulatory authority, the purpose of renewing a license is to maintain the availability of the nuclear facility to meet energy requirements beyond the current term of the facility's license. Thus, whether the facility will continue to operate is based on factors such as the need for power or other matters within the state's jurisdiction or the financial interests of the owners.

1.2.11 What other types of licenses related to reactors does NRC review?

The NRC reviews and grants initial licenses and renewal of licenses for research reactors, and independent spent fuel storage installations (ISFSIs). The NRC also reviews applications for licenses at facilities related to the production of fuel for nuclear reactors or the storage or disposal of waste from reactor operations. These facilities include those that possess and use special

NRC Offices in Rockville, Maryland

nuclear material for uranium milling or production of uranium hexafluoride, fuel fabrication, scrap recovery and waste storage and disposal facilities that are located away from power reactor facilities.

1.3 Timing and Scheduling of a License Renewal review

1.3.1 How does a licensee begin the license renewal process?

The renewal application is the principal document that an applicant provides to both request and support license renewal. The license renewal application includes both general and technical information that

A nuclear power plant licensee may apply to the NRC to renew a license as early as 20 years before expiration of the current license.

demonstrates that an applicant is in compliance with the NRC's regulations in 10 CFR Part 54. An applicant must analyze the aging-related issues for certain passive long-lived structures, systems and components at the facility during the period of the renewed license and describe how these issues will be managed or mitigated. This information must be sufficiently detailed in the application to permit the NRC staff to determine if the applicant's management of these issues is adequate to allow operation during the extended period of operation without undue risk to the public and workers' health and safety. The applicant must also prepare an evaluation of the potential impacts to the environment of facility operation for an additional 20 years.

1.3.2 How early can a licensee request license renewal?

According to the regulations a nuclear power plant licensee may apply to the NRC to renew a license as early as 20 years before expiration of the current license. The NRC staff has determined that 20 years of operating experience is sufficient to assess aging and environmental issues at the site. A licensee may submit application for license renewal at a plant that has less than 20 years of operating experience; however, an exemption to the regulations is required.

It typically takes 22 to 30 months for the NRC to complete its review of a license renewal application and grant the renewed license.

1.3.3 Why does a licensee request license renewal so far in advance of the expiration date of the current license?

A major consideration for seeking license renewal so far in advance of the expiration date of the current license is that it takes about 10 years to design and construct major new generating facilities and long lead times are required by energy-planning decision-makers.

1.3.4 How close to the end of a license can a licensee request that a license be renewed?

License renewal applicants are expected to apply at least 5 years before their license expires. Typically it takes 22 to 30 months (depending on whether there is a hearing) for the NRC to determine whether or not to grant the renewed license. If an applicant submits its application less than 30 months prior to the expiration date of the current license, it is possible that the operating license could expire before the license is renewed. There is a provision in 10 CFR Part 2 that states, "if a licensee files a sufficient application for renewal of an operating license at least 5 years prior to the expiration of the existing license, the existing license will not be deemed to have expired until the application has been finally determined." This provides the applicant the assurance that the facility can continue to operate until the NRC makes a final decision on the license renewal application. The Commission may grant an exemption to the regulations to allow a licensee with less than 5 years remaining on its operating license to submit a license renewal application and still avail itself of the timely application to renew provisions of 10 CFR 2.209(b).

> *NRC's review consists of a*
> - *safety review*
> - *environmental review*
> - *inspections*
> - *independent review by ACRS*

1.3.5 What does the NRC's review consist of?

NRC's review of an application for license renewal has four components: a safety review, an environmental review, inspections, and an independent review by the Advisory Committee on Reactor Safeguards (ACRS). A flowchart of the license renewal process is shown in Figure 1.1.

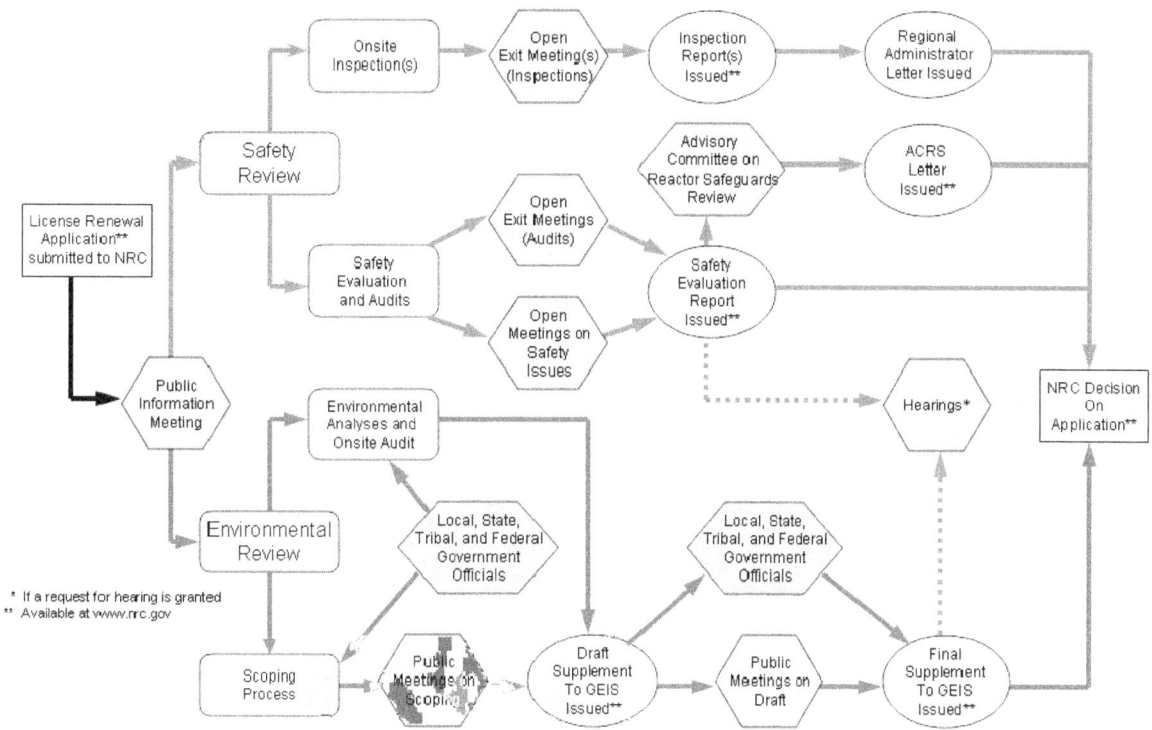

Figure 1.1. License Renewal Process

The NRC staff performs a safety review of the information provided in the application (as supplemented with additional information provided by the applicant at the NRC's request). The results of the staff's safety review are documented in a publicly available safety evaluation report.

The NRC staff's environmental review results in the publication of a publicly available site-specific draft SEIS on license renewal. The public is invited to comment on the draft SEIS. Then, after considering all public comments, the NRC staff issues the final SEIS.

Teams of inspectors with experience in nuclear plant safety visit the site and verify that the applicant has implemented its aging management plans as committed to in the application. The results of plant inspections conducted as part of the license renewal are documented in inspection reports and are made publicly available. The results are also included in the safety evaluation report.

The ACRS is an independent panel of experts that advises the Commission on matters related to nuclear safety. The ACRS reviews the applicant's safety analysis report, the staff's safety evaluation report, and the results of the onsite inspections and makes its recommendation to the Commission regarding issuance of the renewed license.

1.3.6 What is the NRC staff's schedule for the review of a license renewal application?

The NRC staff has developed a review schedule for license renewal. Figure 1.2 shows a generalized timeline of a license renewal review. If a hearing is required, the NRC staff would expect to complete its review of the application within 30 months after receiving the license renewal application; if no hearing is required, the review would be completed within 22 months after receiving the application.

1.3.7 After the current license expires, is the applicant given a new license or is it an extension of the existing license?

At the time of license renewal, the licensee is issued a new license that incorporates and supersedes the existing license. The new license has a new expiration date, which is up to twenty years past the expiration date of the original operating license. The renewed license may also include any license conditions that are specific to license renewal issues.

1.3.8 What if the current operating license has not yet expired? Does the licensee get that time back also or does the renewed license just run for 20 years?

The renewed license is issued for a fixed period of time. That fixed period is the sum of the additional amount of time the applicant requests for license renewal (not to exceed 20 years) plus the remaining number of years on the operating license currently in effect up to a maximum of 20 years. In other words, if a license renewal is granted, the plant will be licensed for the remainder of the time on its current license (20 years or less) plus the renewal time of up to 20 years. In no case, however, may the term of the license, which includes the remaining portion of the original license plus the extension, exceed 40 years.

1.3.9 What happens if the review of a license renewal application is not completed before the current license expires?

If the applicant files a sufficiently complete application for renewal of its operating license at least 5 years before the existing license expires but the renewal of the license is delayed because of administrative or judicial appeal, then the existing license will still be considered valid until a final decision on the renewal application has been made. The Commission may grant an exemption to the regulations to allow a licensee to submit a license renewal application and still avail itself of the timely application to renew provisions of 10 CFR 2.209(b).

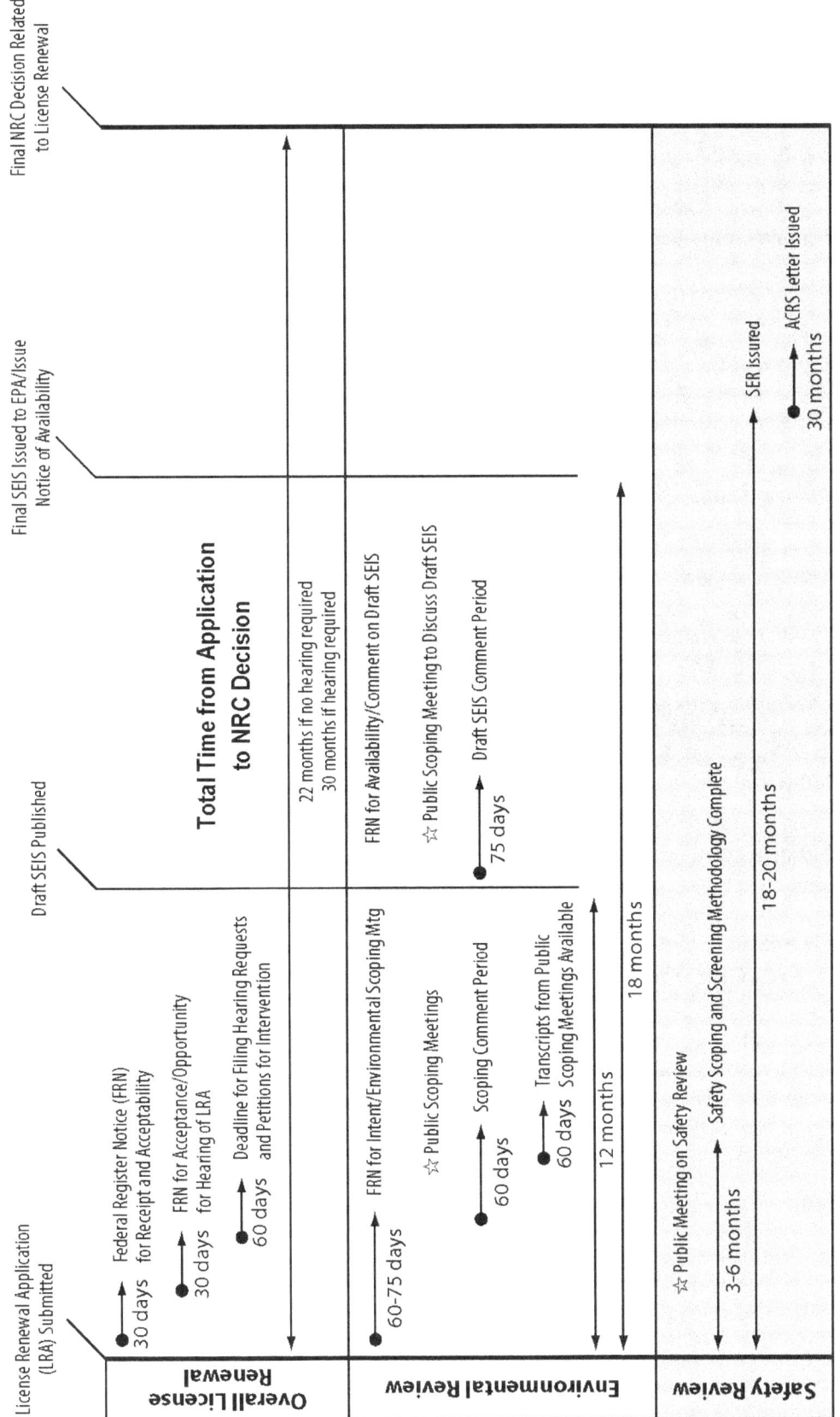

Figure 1.2. Timeline of a License Renewal Review

1.3.10 How many times can a license be renewed?

There are no specific limitations in the Atomic Energy Act or the NRC's regulations restricting the number of times a license may be renewed. However, an applicant has to meet all of the applicable requirements for each subsequent renewal. Any subsequent renewal would require a review similar to that required for the first renewal.

1.3.11 What happens if the licensee sells the facility after the license is renewed?

An NRC regulation (10 CFR 50.80) requires that licenses granted by the NRC shall not be "transferred, assigned, or in any manner disposed of, either voluntarily or involuntarily, directly or indirectly, through transfer of control of the license to any person" unless the Commission gives its consent in writing. Typical staff review of such applications, characterized as requests for restructuring and organizational change, largely consists of ensuring that the ultimately licensed entity has the capability to safely operate the facility, meet the financial qualifications for continued safe operation of the facility and fund the facility decommissioning as required by NRC regulations.

1.3.12 Who pays for the license renewal application and review?

The applicant pays for the preparation of the application. Once the application is submitted to the NRC, the NRC recovers a fee for resources expended in the review. Funding for the NRC is provided by Congress through annual appropriations. Applicants pay fees to the U.S. Treasury to reimburse the government for the cost of the review. Thus, the costs of the development of the license renewal application and the costs of the review are paid for by the applicant and ultimately by electricity consumers.

The costs of the development of the license renewal application and the costs of the review are paid for by the applicant and ultimately by electricity consumers.

1.3.13 If a nuclear facility is relicensed, does that give the applicant a larger window of opportunity to build a new reactor?

Each reactor unit at a commercial nuclear power facility requires a separate operating license. The renewal of an operating license for a specific unit or for all the units at a current site does not permit the applicant to build another reactor unit at the site. If an applicant wants to build a new reactor unit, the applicant must submit an application for a new construction permit or combined operating license.

2.0   Regulatory Basis of the Review

This section discusses the regulatory basis for the review of a license renewal application. The regulations on which the license renewal process is based were established by a rulemaking process. The regulations have been codified and are accessible to members of the public. The NRC uses these regulations as the basis of its review. The regulations are the basis for several other documents, such as review plans or inspection procedures, which specify how the NRC review is conducted.

2.1   How were the regulations for license renewal developed?

The regulations for license renewal were developed through a rulemaking process. The rulemaking process for license renewal started in the early 1980s when the NRC staff recognized that it needed to identify the information required and the process to be used to determine whether to grant renewed licenses for nuclear power reactors. The staff recognized this need because the Atomic Energy Act of 1954 specified that licenses for commercial nuclear reactor facilities would be for 40 years and could be renewed for an additional period of time.

In 1982, the NRC established a comprehensive program for Nuclear Plant Aging Research as the result of a widely attended workshop on nuclear power plant aging. Based on the results of that research, a technical review group concluded that many aging phenomena were readily manageable and did not pose technical issues that would preclude life extension for nuclear power plants.

The NRC also concluded that the existing regulatory requirements governing a nuclear reactor facility would offer reasonable assurance of adequate protection if the license were renewed, provided that the current licensing basis (see response to Question 2.8 for definition of current licensing basis) was modified to account for age-related safety issues. In 1991, the Commission approved a rule on the technical requirements for license renewal and published the rule in the *Code of Federal Regulations*, 10 CFR Part 54. The NRC then undertook a demonstration program to apply the rule to pilot plants and to develop experience to establish implementation guidance. The rule defined the scope as age-related degradation unique to license renewal. However, during the demonstration program, the NRC found that many aging effects are managed adequately during the initial license period. In addition, the NRC found that the review did not allow sufficient credit for existing programs, particularly the maintenance rule, which also helps manage plant-aging phenomena.

As a result, in 1995, following the rulemaking process, the NRC amended the license renewal rule. The amended rule in 10 CFR Part 54 established a regulatory process that is more effective, stable and predictable than the previous license renewal rule. In particular, Part 54 was clarified to focus on managing the adverse effects of aging. The rule changes were intended to ensure that important systems, structures, and components would continue to perform their intended function during the 20-year period of extended operation.

The Commission determined that the NRC would prepare an environmental impact statement for every nuclear power plant license renewal decision to fulfill its responsibilities under the National Environmental Policy Act (NEPA). In parallel with rulemaking on nuclear plant aging, the NRC pursued a separate rulemaking, 10 CFR Part 51, that focused on environmental issues. The final rule, which revised 10 CFR Part 51 and addressed environmental impacts of license renewal, was published in June 1996.

2.2   How does the rulemaking process work?

The process of developing NRC's regulations is called rulemaking. Usually, the NRC's technical staff initiates a change to a regulation because of a safety or environmental concern, an improvement in technical understanding, or an improvement in the regulatory process. The Commission may also direct changes to the regulations. However, any member of the public may petition the NRC under 10 CFR 2.802 of the NRC's regulations to develop, change, or rescind a regulation.

The rulemaking process has several steps. In a rulemaking initiated by the NRC, a rule is proposed after an NRC decision on the need for and general framework of a rule is made. The proposed rule is published in the *Federal Register* and usually contains background information, an address for submitting comments, a date by which comments should be received in order to guarantee consideration by the staff, an explanation of why the regulation change is thought to be needed, and the proposed changes to the text of the regulations. Usually, the public is given 75 to 90 days to provide written comments. Once the public comment period is closed, the staff analyzes the comments, makes any needed revisions to the proposed rule, and forwards the final rule for the Commission's approval, signature, and publication in the *Federal Register*. Each final rule that involves significant matters of policy is sent to the NRC Commissioners for approval. Once approved, the final rule is published in the *Federal Register* and usually becomes effective 30 days after publication.

Not all rules are issued for public comment. Generally, those not published for public comment pertain to NRC organization, procedures, or practices, are interpretations of rules, or are rules for which delaying their publication in order to receive comments would be contrary to the public interest or impracticable.

2.3   Where are the regulations for license renewal found?

The regulations for license renewal are initially published in the *Federal Register* and then included in the *Code of Federal Regulations* (CFR). The CFR is a codification of the general and permanent rules published in the *Federal Register* by the executive departments and agencies of the Federal Government. It is divided into 50 "titles", which represent broad areas subject to Federal regulation. Each title is divided into chapters, which usually bear the name of the issuing agency. The NRC's regulations are found in Title 10 (10 CFR). Each chapter is further subdivided into "Parts" covering specific regulatory areas.

The regulations related to the renewal of licenses are found in 10 CFR Part 51, "Environmental Protection Regulations for Domestic Licensing and Related

Regulatory Functions," and in 10 CFR Part 54, "Requirements for Renewal of Operating Licenses for Nuclear Power Plants." The response to Question 5.2.6 provides information on obtaining a copy of the relevant sections of the *Code of Federal Regulations*.

2.4 How does the NRC conduct its review of a license renewal application?

For each application, the license renewal process includes two reviews: an environmental review and a safety review. The NRC regulations covering these reviews are found in 10 CFR Part 51 and 10 CFR Part 54, respectively. The NRC has issued two sets of regulatory documents (10 CFR Part 51 for environmental issues and 10 CFR Part 54 for safety issues) that describe the NRC's expectation for the format and content of license renewal applications as well as the methods used by NRC staff in evaluating these applications. When an applicant submits a license renewal application to the NRC, the application must contain technical information and evaluations of the environmental and safety issues discussed in the NRC's guidance documents. The NRC reviews the information submitted in the application and requests additional information from the applicant as needed. The NRC teams (comprising NRC staff and contractor personnel) visit the site to conduct audits of environmental and safety records, to conduct interviews with offsite and licensee representatives, to observe operating practices, and to develop an independent assessment. The environmental review also includes an opportunity for input from the public. Given this information, the NRC staff determines whether there is reasonable assurance that the plant can be operated during the period of extended operation without undue risk to the health and safety of the public and to the environment.

Additional details related to the safety and environmental reviews are given in Sections 3.0 and 4.0, respectively, of this publication.

2.5 What defines NRC's review of a license renewal application?

The NRC's review is based on the regulations published in the *Code of Federal Regulations* (10 CFR Parts 51 and 54); however, the NRC provides guidance for the information needed in the applications and for the methods used to conduct the review in sets of NRC documents for safety and environmental issues:

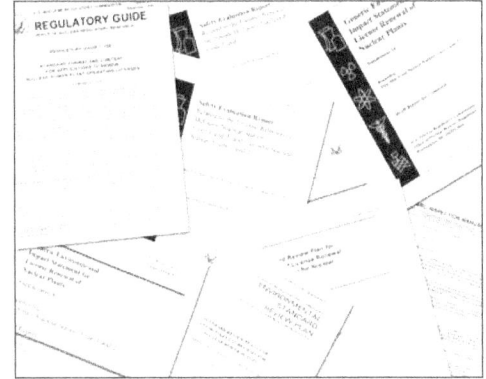

NRC Guidance Documents

Safety Review

- *Code of Federal Regulations* – The scope of the safety review is based on the regulations provided in 10 CFR Part 54, "Requirements for Renewal of Operating Licenses for Nuclear Power Plants."

- *Standard Format and Content for Applications to Renew Nuclear Power Plant Operating Licenses* (Regulatory Guide 1.188) – This document outlines the format and content to be used by the applicant to discuss the safety-related aspects of its license renewal application. It defines the information the licensee must include in the application, which the NRC staff then reviews.

- *Standard Review Plan for Review of License Renewal Applications for Nuclear Power Plants* (NUREG-1800) – This is the outline for the NRC's review of the safety-related issues. The safety review results in a safety evaluation report that is made available for public review.

- Inspection manual chapters (MCs), inspection procedures (IPs) and temporary instructions (TIs) – MCs, IPs and TIs were written for the NRC staff; they guide the staff in conducting inspections to ensure that licensees meet the NRC's regulatory requirements. For example, the IP, *License Renewal Inspections*, 71002, provides the procedures for inspecting and verifying the documentation, implementation, and effectiveness of the programs and activities associated with an applicant's license renewal program. *Policy and Guidance for License Renewal Inspection Programs*, MC-2516, provides guidance to NRC staff for review and inspection activities.

Environmental Review

- *Code of Federal Regulations* – The scope of the environmental review is based on the regulations provided in 10 CFR Part 51, "Environmental Protection Regulations for Domestic Licensing and Related Regulatory Functions."

- *Preparation of Supplemental Environmental Reports for Applications to Renew Nuclear Power Plant Operating Licenses* (Supplement 1 to Regulatory Guide 4.2) – This document outlines the format and content to be used by the applicant to discuss the environmental aspects of its license renewal application. It defines the information the applicant must put in the application, which the NRC staff then reviews.

- *Standard Review Plan for Environmental Reviews for Nuclear Power Plants – Supplement 1: Operating License Renewal* (NUREG-1555, Supplement 1) – This is the outline for the NRC's review of the environmental issues. The review results in a site-specific supplement to the *Generic Environmental Impact Statement for License Renewal of Nuclear Plants*, NUREG-1437 (GEIS).

The NRC has established a license renewal process with clear requirements, to assure safe and environmentally sound operation of the facility for the extended license period.

2.6   Can the NRC change the way it conducts its review or conduct its review in an *ad hoc* manner?

The NRC does not conduct its review in an *ad hoc* manner. The NRC has established a license renewal process with clear requirements, which are codified in 10 CFR Part 51 and 10 CFR Part 54. The process and requirements were developed to assure safe and environmentally sound operation of the facility for the extended license period. In addition, as a basis for the review, the NRC staff uses the regulatory documents discussed in the response to Question 2.5 (including two standard review plans), which describe the methods acceptable to the NRC staff for implementing the license renewal process and the techniques used by the staff in evaluating applications for license renewals.

The NRC continually evaluates the process that is being used and any lessons learned from conducting reviews of applications to renew licenses. It also accepts comments and recommendations from industry and the public related to problems or improvements that could be made. If it appears that there is a need to change the regulations, they can be revised using a rulemaking process, which is described in the response to Question 2.2. Changes that are made in regulatory guidance (for example, regulatory guides and standard review plans) are published for public comment as well.

2.7 What if you disagree with the regulations? How do you get the regulations changed?

The NRC welcomes public participation in the rulemaking process. There are several ways for the public to participate in rulemaking:

- The public may provide comments in response to *Federal Register* notices. The NRC publishes notices of rulemaking activities in the *Federal Register* to solicit public comment, and may also publish a notice of a meeting or workshop to be held regarding a rule. The *Federal Register* notice contains information on how to provide specific comments on a proposed rule to the NRC.

- The public may provide comments on the NRC's RuleForum website. NRC's RuleForum is a web-based computer forum that was developed to provide an easy means for a member of the public to access and comment on NRC rulemaking activities. RuleForum contains proposed rulemakings that have been published by the NRC in the *Federal Register*, petitions for rulemakings that have been received and docketed by the NRC, and other types of documents related to rulemaking.

- Members of the public can provide comments on the NRC's Technical Conference Forum website. The Technical Conference Forum is a web-based forum that facilitates public participation on NRC issues related to the development of draft rulemakings, draft guidance documents, and other initiatives.

- Members of the public may petition the NRC to develop, change or rescind a rule by filing a petition for rulemaking in accordance with the regulations in 10 CFR 2.802.

The NRC welcomes public participation in the rulemaking process.

Before filing a petition for rulemaking, a member of the public may consult with the NRC concerning questions about NRC regulations by calling the Rules and Directives Branch at 301-415-7163 or toll-free at 800-368-5642 or by writing to the following address:

Chief
Rules and Directives Branch
Division of Administrative Services
Office of Administration
U.S. Nuclear Regulatory Commission
Washington, DC 20555-0001

The information that members of the public can receive when consulting with the NRC about a petition for rulemaking includes a description of the procedures and process for filing and responding to a petition for rulemaking, clarification of an existing NRC regulation and the basis for the regulation, or assistance in clarifying their potential petition so that the Commission is able to understand the nature of the issues that are of concern.

Petitions should be submitted to the following address:

Secretary
U.S. Nuclear Regulatory Commission

Washington, DC 20555-0001
Attn: Rulemakings and Adjudications Staff
E-mail: secy@nrc.gov
Fax: 301 415-1101

The petition must, as a minimum, outline a general solution to a problem or present the substance or text of any proposed regulation or amendment or specify the regulation that the petitioner proposes to be rescinded or amended. In writing a petition, a member of the public should state clearly and concisely his or her grounds for and interest in the proposal and include a statement in support of the petition that outlines the specific issues involved; the views or arguments regarding those issues; the relevant technical, scientific, or other data that is reasonably available; and any other pertinent information to support the proposal.

2.8   What is the current licensing basis and how is it applied to license renewal?

The current licensing basis is the particular set of NRC requirements applicable to a licensed operating nuclear power facility.

The current licensing basis (CLB) is the particular set of NRC requirements applicable to a licensed operating nuclear power facility. An applicant for license renewal is also the licensee for the licensed operating nuclear power plant. The CLB includes the applicant's written regulatory commitments for ensuring compliance with and operation within the applicable NRC requirements and the plant-specific design basis. Documents that are in the CLB include:

- NRC regulations contained in applicable parts of Title 10 of the *Code of Federal Regulations* (specifically Parts 2, 19, 20, 21, 26, 30, 40, 50, 51, 54, 55, 70, 72, 73, and 100) and associated appendices,

- NRC Orders,

- Safety and environmental license conditions,

- Technical specifications and environmental protection plans,

- Exemptions,

- Plant-specific design information, as documented in the most recent final safety analysis report (FSAR), and

- NRC environmental reviews (EISs, supplements, and environmental assessments), and

- The licensee's commitments remaining in effect that were made in docketed licensing correspondence, such as responses to NRC bulletins, generic letters and enforcement actions, NRC safety evaluations or licensee event reports.

The CLB changes as documents such as the FSAR or the Technical Specifications are revised or as the licensee's regulatory commitments change. As a result, the NRC requires that each year after submittal of the license renewal application and at least 3 months before scheduled completion of the NRC review, the applicant submit an amendment to the renewal application that identifies any change to the CLB of the facility that would materially affect the contents of the license renewal application.

3.0   NRC Safety Review

The NRC performs a safety review of the applicant's license renewal application to determine if the applicant has adequately demonstrated that the effects of aging will not have adverse impacts on the nuclear facility's operation. This section answers the key questions that are asked related to why this safety review is performed, how it is conducted, and what type of public involvement occurs as a part of the safety review process.

3.1   Why does the NRC perform a safety review?

The NRC performs a safety review to determine whether there is reasonable assurance that activities authorized by the renewed license will continue to be conducted in accordance with the current licensing basis (see response to Question 2.8).

The intent of the NRC's safety review is to determine if the applicant has adequately demonstrated that the effects of aging will not adversely affect any systems, structures, or components, as identified in 10 CFR 54.4. When the plant was designed, certain assumptions were made about the length of time the plant would be operated. During the renewal process, the applicant must also confirm whether these design assumptions will continue to be valid throughout the period of extended operation or whether aging effects will be adequately managed. The applicant must demonstrate that the effects of aging will be managed in such a way that the intended functions of "passive" or "long-lived" structures and components (such as the reactor vessel, reactor coolant system, piping, steam generators, pressurizer, pump casings, and valves) will be maintained during extended operation. For active components (such as motors, diesel generators, cooling fans, batteries, relays, and switches) surveillance and maintenance programs will continue throughout the period of extended operation.

The NRC performs a safety review to determine whether there is reasonable assurance that activities authorized by the renewed license will continue to be conducted in accordance with the current licensing basis.

If additional aging management activities are needed, the applicant may be required to establish new monitoring programs or increase inspections. For instance, applicants should specify activities that need to be performed (such as water chemistry and inspections) to prevent and mitigate age-related degradation. These activities increase the likelihood that the program is effective in minimizing degradation and that a component is replaced if specified thresholds are exceeded.

3.2   What is the basis for the NRC's safety review?

The regulations in 10 CFR Part 54 provide the basis for the NRC's safety review. Detailed guidance on the NRC's safety review for license renewal is provided in the *Standard Review Plan for Review of License Renewal Applications for Nuclear Power Plants* (NUREG-1800). The purpose of the *Standard Review Plan* is to ensure quality and uniformity in staff reviews and to present a well-defined basis from which to evaluate the applicant's programs and activities for the period of extended operation. The *Standard Review Plan* was based on information developed in the *Generic Aging Lessons Learned (GALL) Report* (NUREG-1801), which was developed by the NRC with input from interested stakeholders. The report documents the basis that is used for determining when existing programs are adequate and when they should be augmented for license renewal.

Spent Fuel Pool

3.3 How is the safety review performed?

The NRC Office of Nuclear Reactor Regulation reviews an applicant's renewal application and supporting documentation based on the *Standard Review Plan for Review of License Renewal Applications for Nuclear Power Plants* (NUREG-1800). Audits are also performed to review onsite documentation supporting the application. In addition, the NRC follows the guidance in the NRC Inspection Manual Chapter (MC)-2516, *Policy and Guidance for License Renewal Inspection Programs*, and the NRC Inspection Manual – Inspection Procedure 71002, *License Renewal Inspections*, which contains policies and guidance for license renewal inspection programs. The license renewal application is reviewed to determine if the applicant meets the technical and regulatory requirements of the regulations. Specifically, the application must identify those systems, structures, and components that are within the scope of license renewal and subject to an aging management review and must also identify applicable aging mechanisms and describe programs in place to manage aging. This review commonly results in the NRC's requesting additional information from the applicant.

The safety review process also includes site inspections to assess whether the applicant has implemented and complied with the regulations for license renewal. The inspection teams are composed of technical, program, and operational experts from the NRC and its consultants. Teams of specialized inspectors travel to the reactor site, normally twice, to verify whether the effects of aging will be managed such that the plant can be operated during the period of extended operation without undue risk to the health and safety of the public. The review results in a publicly available safety evaluation report.

3.4 How is the scope of the safety review for license renewal determined?

Turbine Deck

The scope of the safety review is based on the regulations provided in 10 CFR Part 54, "Requirements for Renewal of Operating Licenses for Nuclear Power Plants." The NRC has also developed other documents to provide greater information on the type of material reviewed and the depth of the review:

- The format and content of the safety aspects of a license renewal application are given in *Standard Format and Content for Applications to Renew Nuclear Power Plant Operating Licenses* (Regulatory Guide 1.188). This document defines the information the applicant must put in the application, which the NRC staff reviews.

- The NRC's review of the safety-related issues is outlined in a Standard Review Plan that was written specifically for the NRC's safety review of license renewal applications, "*Standard Review Plan for*

Review of License Renewal Applications for Nuclear Power Plants" (NUREG-1800). The review results in a safety evaluation report that is made available to the public.

- The staff is guided in conducting inspections by the NRC Inspection Manual – Manual Chapter (MC)-2516, *Policy and Guidance for the License Renewal Inspection Programs* and the NRC Inspection Manual – Inspection Procedure 71002, *License Renewal Inspections*. MC-2516 provides guidance to NRC staff and consultant personnel for review and inspection activities associated with an applicant's license renewal program. Inspection Procedure 71002 provides the procedures for inspecting and verifying the documentation, implementation, and effectiveness of the programs and activities associated with an applicant's license renewal program.

3.5   Why is NRC focusing its safety review on aging-management issues?

The focus of the license renewal safety review is on managing the detrimental effects of aging. The review provides reasonable assurance that the effects of aging will be managed for the period of extended operation such that systems, structures, and components (SSCs) will continue to perform their intended functions in accordance with the plant's current licensing basis. Many of the existing programs and regulatory requirements that already provide adequate aging management will continue to be applicable after renewal. The license renewal review focuses on those SSCs for which current activities and requirements may not be sufficient to manage aging in the period of extended operation.

3.6   What is embrittlement?

Embrittlement is an aging process that makes material more brittle and more susceptible to fractures. There are two processes that occur in a nuclear reactor that cause embrittlement of metals. One is continual irradiation of materials by neutrons. This occurs in components like the reactor vessel or reactor coolant system where the neutron flux is the greatest. The second is thermal aging embrittlement from wide temperature fluctuations that occur in the structures and components associated with the production of steam. Thermal aging embrittlement is known to occur in cast austenitic stainless steel, which is a nonmagnetic stainless steel used in various parts of the facility, including the piping, pump casings, and reactor vessel internals.

3.7   Is reactor embrittlement being reviewed during license renewal?

Reactor Vessel

Yes, reactor embrittlement is being reviewed during the license renewal process. The Commission requires that an applicant detect and mitigate the effects of aging, beginning with an examination and verification that the systems, structures, or components function as they were originally intended to and that their functions have not been compromised or degraded.

3.8 Is reactor vessel head degradation like the problem at Davis-Besse Nuclear Power Station being considered during license renewal?

Replacement Reactor Vessel Head for R.E. Ginna Nuclear Power Plant post-Davis-Besse

The reactor vessel head corrosion event at the Davis-Besse Nuclear Plant is an operational issue and outside the scope of license renewal. The event has had, and continues to have, a significant effect on both the NRC and reactor licensees. The corrosion event was unexpected and unknown to both the industry and staff. The corrosion was discovered by the licensee during an NRC-required inspection resulting from safety concerns related to reactor vessel head nozzle circumferential cracking. Since the discovery of the reactor vessel head corrosion event at Davis-Besse, the NRC has significantly increased the oversight of licensee reactor vessel head activities and other activities that may affect the condition of the reactor vessel head. Almost immediately after the discovery, the NRC strengthened reactor vessel head inspections by the imposition of inspection requirements by order. The immediate initiatives by the NRC staff provide assurance that any further corrosion events will be identified early and corrected. The NRC also formed a Lessons Learned Task Force (LLTF) to carefully review the Davis-Besse incident and make recommendations for improvement. The LLTF has made recommendations for improvements in reactor vessel inspection requirements, inspection program management and inspector qualification, handling of operating experience information, and research activities relating to leakage detection methodologies. Forty-nine recommendations arising from the LLTF review were adopted for implementation by the NRC staff, and over 40 have already been implemented. All but one, dealing with updating the American Society of Mechanical Engineers (ASME) code, have been completed. The progress of implementing the 49 recommendations is reported semiannually to the NRC's Executive Director of Operations and then forwarded to the NRC Commissioners. The NRC is confident that the implementation of the 49 LLTF recommendations will preclude any future recurrence of reactor vessel head corrosion similar to that at Davis-Besse.

3.9 What sort of inspections are conducted at the plant?

Facility Inspection

The NRC maintains an inspection program for operating nuclear power facilities. The NRC inspection program assesses whether activities are properly conducted and equipment is properly maintained to ensure safe operations. The NRC inspection program is continuous and relies primarily on resident inspectors, who are stationed at each nuclear reactor facility, and region-based inspectors, who supplement the activities of the resident inspectors.

In addition to the inspection program for operations, the NRC has established an inspection program for license renewal that examines the information provided by the applicant in the renewal application. The NRC reviews the license renewal application to determine if the applicant has identified those systems, structures, and components that must be in the scope of license renewal, has identified applicable aging effects, and has programs in place to manage aging. The site inspections are assessments of the applicant's implementation of and compliance with the regulations in 10 CFR Part 54. The inspection team includes technical, program, and operational experts from the NRC and its consultants. The intent of the inspections is to determine whether the effects of aging will be managed such that the facility can be operated during the period of extended operation without undue risk to the health and safety of the public and to ensure the consistency of the applicant's programs to manage aging within the current licensing basis.

3.10 What documents are generated during the NRC staff's review of the license renewal application?

There are several documents generated during the NRC's review of a license renewal application:

- The safety evaluation report (SER) documents the results of the NRC staff's review of aging-management and the applicant's programs to address these matters during the period of extended operation.

NRC Review Reports

- The environmental impact statement (EIS) in the form of a site-specific supplement to the generic environmental impact statement (SEIS) documents the results of the NRC staff's review of the potential environmental impacts of continued operation of the plant during the period of extended operation.

- Inspection reports document the results of the NRC staff's inspections of the applicant's implementation of its quality assurance program and aging-management programs.

- The letter from the Advisory Committee on Reactor Safeguards (ACRS) to the Commission documents the results of the ACRS independent review of the safety aspects of the license renewal application and the staff's SER.

- If there is a hearing, the initial decision by the Atomic Safety and Licensing Board (ASLB) documents the findings of the ASLB on those items that were litigated in the hearing.

- A Commission paper, issued by the Executive Director for Operations (EDO), summarizes the conclusions documented in all of the above documents, and provides the NRC staff's recommendation concerning whether the operating license should be renewed. For uncontested applications, an EDO memorandum summarizes the staff's decision on the renewal of the license.

The NRC staff documents the results of its review of the license renewal application in a safety evaluation report. The results of the staff's safety review are available to the public.

3.11 Who uses the safety review documents (such as safety evaluation reports) that NRC publishes?

The NRC staff documents the results of its review of the license renewal application, first, in a draft and, then, in a final safety evaluation report (SER). When it is initially issued, the draft SER identifies any remaining open and confirmatory items that the staff is still resolving with the applicant. The final SER is used as one input into the decision as to whether to renew the license. The SER is considered during the independent review of the application by the Advisory Committee on Reactor Safeguards (ACRS). If there is a hearing, the SER provides the basis for the NRC staff's positions and is considered by the Atomic Safety and Licensing Board (ASLB) during the adjudicatory process. Finally, the SER is considered by the Director of the Office of Nuclear Reactor Regulation and, if the application is contested, by the Commission when the agency decides whether to renew the license.

3.12 Is the public provided the opportunity to comment on the results of the NRC staff's safety review?

During the safety review process, the staff holds meetings with the applicant to discuss the review of the application. The public is invited to observe and has the opportunity to comment at the conclusion of the technical portion of the meeting.

The results of the staff's safety review are available to the public. However, the highly technical nature of the staff's safety review does not lend itself to a public involvement process such as that used for the environmental review. As a result, there is no notification in *Federal Register* notices related to an opportunity to comment on the safety review prior to its issuance. However, a draft Safety Evaluation Report is available electronically from the Publicly Available Records System (PARS) component of the NRC's Agency-wide Documents Access and Management System (ADAMS). The ADAMS Public Electronic Reading Room is accessible from the NRC website at http://www.nrc.gov/NRC/ADAMS/ index.html. Additionally, the public can provide comments to the Advisory Committee on Reactor Safeguards (ACRS) on the staff's review of the license renewal application in advance of the ACRS meeting.

In addition, any person who believes he or she would be adversely affected by a specific reactor license renewal may request a hearing. Members of the public may also petition the Commission, in accordance with the provisions of 10 CFR 2.206, for consideration of safety issues during current operation and the period of extended operation of the plant.

3.13 When can members of the public bring up safety issues that should be considered during the license renewal review?

System Inspection

Although there is not a formal public comment period for the safety review, members of the public can raise safety issues related to license renewal to the attention of the NRC staff during the license renewal review period. Members of the public may raise issues during the review process in public meetings or directly to the NRC project managers for the license renewal process. Members of the public can also bring safety issues to the NRC's attention by emailing or phoning directly, as discussed in the response to Question 5.1.11. If the issue is one that affects the current operation of a facility, the NRC assesses the issue and, if needed, requires a licensee to take appropriate action apart from the license renewal evaluation process.

Members of the public who believe they would be adversely affected by the renewal have an opportunity to request a formal adjudicatory hearing. The process for requesting a hearing is discussed in the response to Question 5.1.9.

3.14 What if, during the course of the safety review, some design flaw or safety problem with the plant is discovered? Does the problem get addressed now or only during the renewal period?

If a design flaw or safety problem affects the plant's current operation, then the issue is addressed immediately rather than waiting for the conclusion of the staff's license renewal review or the beginning of the license renewal period.

3.15 Does the license renewal process result in any changes to the license conditions imposed on the applicant?

The license renewal process will result in changes to the current licensing basis for the facility (see response to Question 2.8 for a definition of current licensing basis). The premise is that the applicant's evaluation process results in a current licensing basis that is adequate to ensure safe operation of the facility during the period of extended operation. However, in cases where the NRC review determines that additional requirements are necessary for safe operation, the NRC will require that additional changes be made to the current licensing basis. These changes may be in the form of a license condition or technical specification requirement. A license condition is a condition that must be met for the license to be valid. A technical specification is a specific requirement that is contained in the license that a licensee must comply with.

3.16 What is the Advisory Committee on Reactor Safeguards and how are they involved in license renewal?

The ACRS is an independent panel of experts that advises the Commission on matters related to nuclear safety.

The Advisory Committee on Reactor Safeguards (ACRS) is an advisory committee mandated by the Atomic Energy Act of 1954, as amended, under the Federal Advisory Committee Act (FACA). The Committee has three primary purposes:

- to review and report on safety studies and reactor facility license and license renewal applications,

- to advise the Commission on the hazards of proposed and existing reactor facilities and the adequacy of proposed reactor safety standards, and

- to initiate reviews of specific generic matters or nuclear facility safety-related items.

The ACRS is independent of the NRC staff and reports directly to the Commission, which appoints its members. The operational practices of the ACRS are governed by the provisions of the FACA. The ACRS is composed of recognized technical experts in their fields. It is structured so that experts representing many technical perspectives can provide independent advice, which can be factored into the Commission's decision-making process. Most Committee meetings are open to the public and any member of the public may request an opportunity to make an oral statement during the committee meeting.

During the license renewal process, the ACRS acts as an independent third-party oversight group that reviews and makes recommendations to the Commission on the safety aspects of renewal applications. The ACRS mandate does not include National Environmental Policy Act (NEPA) reviews.

4.0 Environmental Review

The NRC performs an environmental review of an applicant's license renewal application to determine the environmental effects of operating the nuclear power facility for an additional 20 years. The Commission determined that the NRC would prepare an environmental impact statement for each license renewal action to fulfill its responsibilities under the National Environmental Policy Act of 1969 (NEPA). NEPA requires that all Federal agencies consider environmental values in the conduct of their work. This section answers key questions related to NEPA and on the environmental review process for license renewal.

The National Environmental Policy Act of 1969 (NEPA) defined a national policy for the environment and established the basis for considering environmental issues in the conduct of Federal activities.

4.1 National Environmental Policy Act (NEPA)

NEPA defined a national policy for the environment and established the basis for considering environmental issues in the conduct of Federal activities. This section describes NEPA, the requirements that are included in the Act, as well as special features of NEPA, including tiering and scoping. This section also includes a discussion of the requirements for Federal agencies to comply with NEPA and the overview process that is built into the NEPA regulations.

4.1.1 What is the National Environmental Policy Act (NEPA)?

NEPA establishes a national policy which:

* encourages productive and enjoyable harmony between man and his environment,

* promotes efforts which will prevent or eliminate damage to the environment and biosphere and stimulate the health and welfare of man, and

* enriches the understanding of the ecological systems and natural resources important to the Nation.

U.S. Capitol

The legal citation for NEPA is 42 U.S.C. 4321 et seq. The text of the National Environmental Policy Act can be found in Public Law 91-190, 42 U.S. Code 4321-4347, January 1, 1970, and in its subsequent amendments. The amendments were PL 94-52, July 3, 1975; PL 94-83, August 9, 1975; and PL 97-258, paragraph 4(b), September 13, 1982.

NEPA also established the President's Council on Environmental Quality (CEQ). On November 29, 1978, the CEQ issued regulations (40 CFR Part 1500) implementing NEPA. These regulations became effective for and binding upon all Federal Executive Branch agencies within a year after the regulations were published. The regulations direct Federal agencies on matters related to environmental policy, including the public scoping process, use of lead agencies, and selection of alternatives. The NRC is an independent regulatory agency. In establishing its own regulations, the Commission has announced its policy to take account of the CEQ's 1978 regulations voluntarily, subject to certain conditions (see 10 CFR 51.10).

4.1.2   What does the National Environmental Policy Act (NEPA) require?

NEPA requires all Federal agencies considering a major Federal action to take the following actions:

- utilize a systematic, interdisciplinary approach for decision-making on actions that may have an impact on the environment,

- inform and involve the public in the decision-making process,

- consider significant environmental impacts associated with the action, including cumulative impacts,

- consider alternatives and their impacts to the proposed action, and

- require a candid discussion and evaluation of impacts and mitigation alternatives.

4.1.3   What is not required by the National Environmental Policy Act (NEPA)?

NEPA does not require the review or re-analysis of actions other than the action being considered.

NEPA does not require that the Federal agency choose the alternative with the least impact but rather that it disclose all potential impacts so that the decision that the agency makes can be fully informed. NEPA does not provide for adjudication of contested actions. Each agency's administrative procedures specify the conditions under which administrative hearings are held.

NEPA does not require the review or re-analysis of actions other than the action being considered. For example, the NEPA review for license renewal would not include an environmental review of the existing operating license, a review of an independent spent fuel storage installation, or an analysis of a waste repository, each of which has had its own separate NEPA review.

4.1.4   What is an environmental impact statement (EIS)?

An EIS is a written analysis of the reasonably foreseeable effects of an activity on the environment.

An EIS is a written analysis of the reasonably foreseeable effects of an activity on the environment, including the air, water, animal life, vegetation, and natural resources, and on any property of historic, archaeological, or architectural significance. The review evaluates cumulative, economic, social (including environmental justice), cultural, and other impacts. The preparation of an EIS includes:

- publication of a notice of intent to prepare the EIS,

- "scoping," that is, preliminary analysis and consultation with other agencies and stakeholders (including the public) to determine the scope of the EIS, defining the range of actions, alternatives, and impacts to be considered,

- analysis leading to a draft EIS,

- public review and comment, responses to the comments, and possibly further analyses, amendments, or revision of the draft EIS, and

- publication of an EIS that includes discussion of the comments made during the public review period.

4.1.5 What is the difference between generic, programmatic, site-specific, and supplemental environmental impact statements?

A generic environmental impact statement (GEIS) is an environmental impact statement (EIS) that assesses the scope and impact of environmental effects that would be associated with an action at numerous sites. For license renewal, the NRC has issued a GEIS, *Generic Environmental Impact Statement for License Renewal of Nuclear Plants*, NUREG-1437. The GEIS assesses the scope and impact of environmental effects at all of the currently existing U.S. nuclear power plant sites. The conclusions of the GEIS provide a list of issues that were analyzed and resolved in a generic fashion and a list of issues that require a site-specific analysis.

A programmatic EIS is an EIS prepared for a broad plan or action that is evolving and includes a number of phases or individual actions. For example, a programmatic EIS was prepared to assess the proposed cleanup of the Three Mile Island Unit 2 reactor after the March 1979 accident.

A site-specific EIS concentrates on proposed activities for a specific geographic location. It may rely on the findings of a GEIS for some or all issues that were determined to be appropriately addressed in a generic fashion. However, for those items requiring a site-specific analysis, the site-specific EIS provides the necessary in-depth assessment required to complete the environmental review for the action. It may consider a single power reactor unit at a specific location, or it may include multiple power reactor units at a specific location.

A supplemental EIS (SEIS) updates or supplements an existing EIS. It is required when the project changes, or if new impacts are discovered after the original EIS is completed. For license renewal, the Commission directed the staff to issue site-specific supplements to NUREG-1437 for each application.

4.1.6 What is tiering?

Tiering is the process of addressing a general program (such as a nuclear power plant license renewal) in a generic (or programmatic) environmental impact statement (EIS), and then analyzing a detailed element of the program (such as a site-specific action related to the general program) as a supplement to the generic EIS. The concept of tiering was promulgated by the President's Council on Environmental Quality (CEQ) in its 1978 regulations implementing the requirements of the National Environmental Policy Act (NEPA). The CEQ has stated that its intent in formalizing the tiering concept was to encourage agencies "to eliminate repetitive discussions and to focus on the actual issues ripe for decisions at each level of environmental review."

Tiering is the process of addressing a general program in a generic EIS, and then analyzing a detailed element of the program as a supplement to the generic EIS.

For license renewal, a site-specific supplement to the generic environmental impact statement (SEIS) contains summaries of issues resolved in the *Generic Environmental Impact Statement for License Renewal of Nuclear Plants*, NUREG-1437 (GEIS). The detailed analyses from the GEIS are incorporated by reference into the SEIS. Issues that were not resolved in the GEIS receive a detailed site-specific analysis in the SEIS. Thus, the supplement does not duplicate material found in the GEIS.

The CEQ noted that tiering can be a useful method of reducing paperwork and duplication and should be viewed as a means of accomplishing NEPA requirements in an efficient manner. The tiering process makes each EIS of greater use and meaning to the public without duplication of the analysis prepared for the previous impact statement.

4.1.7   What is scoping?

Scoping is one of the steps in preparing an environmental impact statement (EIS) (see response to Question 4.1.4). The President's Council on Environmental Quality's (CEQ's) regulations direct agencies to engage in a scoping process. The purpose of this process is to determine the range of actions, alternatives, and impacts to be considered in the EIS. Scoping is intended to ensure that problems are identified early and are properly studied, that issues of little significance do not consume time and effort, that the draft EIS is thorough and balanced, and that delays occasioned by an inadequate draft EIS are avoided. The scoping process should:

- identify the public and agency concerns,

- clearly define the environmental issues and alternatives to be examined in the EIS, including the elimination of issues that are not significant,

- identify related issues that originate from separate legislation, regulation, or Executive Order, and

- identify state, Tribal, and local agency requirements that must be addressed.

An effective scoping process can help reduce unnecessary paperwork and time delays in preparing and processing the EIS by clearly identifying all relevant issues and procedural requirements.

Public meetings during scoping are not required by the National Environmental Policy Act (NEPA). Instead, the manner in which public input will be sought is left to the discretion of the agency. For license renewal, the NRC has elected to conduct public meetings as a part of the scoping process. These meetings are held in the vicinity of the power reactor facility early in the assessment process. The public is invited to attend the meetings to provide its insights on the scope of the environmental assessment.

4.1.8   How does the NRC implement the National Environmental Policy Act (NEPA)?

NEPA is implemented in the NRC's regulations in 10 CFR Part 51, "Environmental Protection Regulations for Domestic Licensing and Related Regulatory Functions." These regulations are used by the NRC as the basis for conducting environmental impact statements or environmental assessments in support of NEPA.

4.1.9   Who oversees the National Environmental Policy Act (NEPA) and ensures that the NRC does an adequate job of meeting the NEPA requirements?

The President's Council on Environmental Quality (CEQ) oversees Federal agencies' implementation of NEPA. This is accomplished through regulations implementing the procedural provisions of NEPA and through interpretation of statutory requirements. The CEQ was established by Congress in 1969 when NEPA was enacted. The Chairman of the CEQ, who is appointed by the President with the advice and consent of the Senate, serves as the President's principal environmental policy advisor.

The requirement to review environmental impact statements (EISs) is the responsibility of the U.S. Environmental Protection Agency (EPA). As a result, EISs prepared by Federal agencies (including the NRC) are reviewed by the EPA. The EPA also maintains a national EIS filing system and publishes weekly notices of EISs available for review and summaries of the EPA's comments on EISs.

4.1.10 Do environmental impact statements (EISs) written by the NRC get reviewed for adequacy by any other governmental agency?

The requirement to review environmental impact statements is the responsibility of the U.S. Environmental Protection Agency.

The U.S. Environmental Protection Agency (EPA) has the responsibility to review EISs that are prepared by other Federal agencies (including the NRC). This review responsibility is a requirement placed on the EPA by the National Environmental Policy Act (NEPA). In addition to reviewing the EIS for adequacy, the EPA also provides the sponsoring agency (in this case, the NRC) with an assessment of each EIS. The assessment is used as a measure of the NRC's adherence to NEPA. Comments are provided by the EPA to the NRC to use as information on future EISs. Additionally, the EPA comments on draft EISs under its statutory areas of responsibility such as clean water and clean air. Other Federal agencies are invited to participate in the scoping process (see response to Question 4.1.7) and are afforded the opportunity to review and comment on the draft EISs.

4.1.11 Apart from the National Environmental Policy Act (NEPA), does the NRC have to comply with other environmental laws, regulations, or Executive Orders?

The NRC has to comply with all applicable Federal environmental laws, regulations, and Executive Orders, including its own regulations (in Title 10 of the *Code of Federal Regulations*) and those promulgated by other Federal agencies, so long as compliance would not be inconsistent with other statutory requirements. Some of the laws, regulations, and Executive Orders that pertain to the license renewal process include the following:

- Endangered Species Act of 1973, with respect to protecting threatened and endangered species and critical habitats, and initiating formal or informal consultation with the U.S. Fish and Wildlife Service and/or National Marine Fisheries Service,

- Federal Water Pollution Control Act of 1972 (commonly called the Clean Water Act), requiring the restoration and maintenance of the chemical, physical, and biological integrity of water resources,

- Fish and Wildlife Coordination Act of 1958, ensuring consideration of fish and wildlife resources in the planning of development projects that affect water resources,

- Migratory Bird Treaty Act, as amended, controlling endangering or taking migratory birds,

- Coastal Zone Management Act of 1972, with respect to natural resources and land or water use of the coastal zone,

- Marine Mammal Protection Act of 1972, requiring the protection of marine mammals,

- Marine Protection, Research and Sanctuaries Act of 1972, controlling the dumping of dredged material into the ocean,

- Rivers and Harbors Appropriations Act of 1899, controlling the deposition of debris in navigable waters, or tributaries to such waters,

- National Historic Preservation Act of 1966, requiring protection and preservation of significant historic properties during construction, refurbishment and operation of the plant,

- Native American Graves Protection and Repatriation Act of 1990, related to disturbance of Native American burial grounds and cultural sites,

- National Electrical Safety Code, regulating shock hazards from transmission lines,

- Executive Order 12898 (59 FR 7629), requiring Federal executive branch agencies to consider environmental justice in minority and low-income populations,

- 40 CFR Part 6, Appendix A, defining procedures on floodplain and wetlands protection,

- 40 CFR Part 122 and Part 124, implementing the National Pollutant Discharge Elimination System (NPDES) permit conditions for discharges including storm-water discharges,

- 40 CFR Part 125, addressing water-quality standards,

- 40 CFR Part 165, controlling the disposal and storage of pesticides,

- 40 CFR Part 403, regulating waste effluents, and

- 40 CFR Part 700 – 716, defining practices and procedures for managing toxic chemicals.

4.1.12 Does the NRC coordinate or consult with other Federal agencies as part of its environmental reviews?

One of the first requirements for developing an environmental impact statement (EIS) is to publish, in the *Federal Register*, a notice of intent to prepare the EIS and conduct scoping. This *Federal Register* notice is a method of alerting other agencies (including other Federal agencies) that may have an interest in participating in the review or wish to participate in the scoping process.

During the analysis and preparation of the draft site-specific supplement to the generic environmental impact statement for license renewal (SEIS), the NRC staff consults with appropriate Federal agencies. The NRC usually contacts the U.S. Fish and Wildlife Service (Department of the Interior) and the National Marine Fisheries Service (Department of Commerce) for environmental issues related to the impact on any threatened or endangered species that may be in the vicinity of the nuclear power facility or to any critical habitat that could be affected by the licensing action (in this case, license renewal). If other agencies have actions or jurisdiction over areas directly related to the review, they would also be contacted directly by the NRC.

In addition to NRC-coordinated consultation, the draft EIS is reviewed by various Federal agencies at their discretion. For example, at the Federal level, the draft SEISs for license renewal are most commonly reviewed by the U.S. Environmental Protection Agency and the U.S. Department of the Interior. The comments from these agencies are considered and included in the final SEIS, as appropriate.

4.2 NRC Environmental Review Process for License Renewal

The environmental review process for license renewal is described below, beginning with a brief overview of the basis for the NRC environmental reviews. It is followed by a history of the environmental reviews and a discussion of the *Generic Environmental Impact Statement for the License Renewal of Nuclear Plants*, NUREG-1437 (GEIS), and its role in conducting the environmental reviews. The relationship between the GEIS and the site-specific supplement to the generic environmental impact statement on license renewal (SEIS) is described. The process used to review the environmental portion of an application for license renewal is also described. Examples of generic and site-specific issues are given. A discussion of the measure of significance for the issues is also provided. The final set of questions and answers in this section relates to the periodic update of the GEIS.

The License Renewal rule does not constitute a major Federal action significantly affecting the quality of the human environment and, therefore, an environmental impact statement is not required. However, the Commission decided that the NRC would prepare a site specific supplement to the generic environmental impact statement on license renewal to ensure that the public had the highest level of participation in and confidence about the NRC's action on a license renewal application.

4.2.1 Basis for the NRC's environmental review for license renewal

4.2.1.1 Why does the NRC conduct an environmental review for license renewal?

Every licensing action for a nuclear power plant is evaluated to determine whether and to what degree an environmental review is required. Some actions are categorically excluded and do not require an environmental review (see 10 CFR 51.22), some require an environmental impact statement (EIS) (see 10 CFR 51.20), and others require an environmental assessment (see 10 CFR 51.21). By law, the National Environmental Policy Act (NEPA) requires an EIS for "major Federal actions." The Commission determined that an EIS should be prepared for each license renewal application even though the Commission determined that license renewal was not a "major Federal action."

4.2.1.2 If license renewal is not a major Federal action, then why did the Commission decide to conduct an environmental review?

A "major Federal action" is defined by the National Environmental Policy Act (NEPA) implementing regulations to be an "action with effects that may be major and which are potentially subject to Federal control and responsibility." According to the Statements of Consideration that accompany the Final Rule on License Renewal (60 FR 22461), published in 1995, the Commission determined that the license renewal rule did not constitute a major Federal action significantly affecting the quality of the human environment and, therefore, an environmental impact statement (EIS) was not required under NEPA regulations. Nevertheless, in response to comments made by the President's Council on Environmental Quality (CEQ), the U.S. Environmental Protection Agency (EPA), a number of state agencies, and members of the public, the Commission decided that the NRC would prepare a site-specific supplement to the generic environmental impact statement on license renewal (SEIS), rather than an environmental assessment (EA) as initially proposed, for each license renewal application. This decision was made to ensure that the public had the highest level of participation in and confidence about the NRC's action on a license renewal application.

The SEIS would be written following a public scoping period for each renewal application, during which time the NRC would request public comments related to new and significant information that might not have been considered in the analysis in the *Generic Environmental Impact Statement for License Renewal of Nuclear Plants*, NUREG-1437 (GEIS). The decision was also made to issue each supplement in draft form for public comment. These decisions were documented in the "Final Rule for Environmental Review for Renewal of Nuclear Power Plant Operating Licenses," printed in the *Federal Register* on June 5, 1996 (61 FR 28467), and are now included in the NRC's regulations in 10 CFR Part 51.

4.2.2 History of environmental reviews for license renewal

4.2.2.1 What is the history of the environmental reviews for license renewal?

In 1986, the NRC initiated a program to develop the license renewal regulations and associated regulatory guidance in anticipation of applications for the renewal of nuclear power plant operating licenses. The Commission decided that, in addition to the development of license renewal regulations focused on the protection of health and safety, an amendment to its environmental protection regulations was also warranted. In November 1989, the NRC held a public workshop on license renewal. One of the sessions was devoted to environmental issues associated with license renewal and the possible value of amending the NRC's environmental regulations in 10 CFR Part 51. On July 23, 1990, the NRC published in the *Federal Register* (55 FR 29967) an advance notice of proposed rulemaking and a notice of intent to prepare a generic environmental impact statement (GEIS). The proposed rule was published in the *Federal Register* on September 17, 1991 (54 FR 29967). After the comment period on the proposed rule, the NRC communicated with the President's Council on Environmental Quality (CEQ) and the U.S. Environmental Protection Agency (EPA) to address their concerns about procedural aspects of the proposed rule. The NRC also had discussions with officials from state agencies about their concerns related to certain features of the proposed rule, including the states' regulatory authority over the need for power and licensee economics. Three regional workshops and another public meeting (specifically to discuss the states' issues) were held in 1994. After considering comments from the workshops and the written comments, the NRC issued a proposed supplement to the proposed rule, which was published on July 25, 1994. Comments were requested on that proposal and the final rule was published in 1996, containing changes to the regulations in 10 CFR Part 51 to define the environmental reviews for license renewal and require a site-specific supplement to the generic environmental impact statement (SEIS) to support a decision on each license renewal application. The SEISs supplement the *Generic Environmental Impact Statement for License Renewal of Nuclear Plants* (GEIS), which was published in 1996 as NUREG-1437. In 1999, the NRC issued Addendum 1 to the GEIS and modified the rule to correct errors and resolve another issue generically. All references to the GEIS include the GEIS and its Addendum 1.

In 2000, the NRC issued the *Standard Review Plan for Environmental Reviews for Nuclear Power Plants – Supplement 1: Operating License Renewal*, NUREG-1555, providing guidance to the staff on how to review the environmental portions of renewal applications. The NRC also issued in 2000 a supplement to Regulatory Guide 4.2, *Preparation of Supplemental Environmental Reports for Applications to Renew Nuclear Power Plant Operating Licenses*, which provides guidance on the format and content of an environmental report to be submitted as part of a license renewal application. As the NRC gains experience from environmental reviews of license renewal applications, it also expects to update this guidance to further improve the process.

In April 1998, Baltimore Gas and Electric became the first licensee to apply for license renewal for its Calvert Cliffs nuclear power reactors on Chesapeake Bay. The SEIS was published in October 1999 and the plant received a renewed license on March 23, 2000. Duke Energy Corporation followed suit in July 1998, when it sought license renewals for its Oconee nuclear units in South Carolina. The Oconee SEIS was published in December 1999, and the plant received its renewed license on May 23, 2000. As of the beginning of 2006, licenses for 22 facilities (for a total of 39 nuclear power reactor units) have been renewed (see Table 1.1).

> *The GEIS examines the reasonably foreseeable range of environmental impacts that could occur as a result of renewing licenses of individual nuclear power plants.*

4.2.2.2   What were the results of the comprehensive environmental review conducted in the 1990s?

The result of the comprehensive environmental review was the *Generic Environmental Impact Statement for License Renewal of Nuclear Plants* (GEIS), which was issued in 1996 as NUREG-1437. Addendum 1 to the GEIS was issued in 1999 to correct errors and to reflect a new analysis of transportation issues that expanded the generic findings about the environmental impacts due to transportation of fuel and waste to and from a single nuclear power plant. Specifically, this amendment revised the categorization of transportation of high level waste from Category 2 (site-specific) to Category 1 (generic). (References in this document to the "GEIS" includes the GEIS and its Addendum 1.)

Development of the GEIS was based on environmental and safety documentation from original licensing proceedings and on information from state and Federal regulatory agencies, the nuclear licensee industry, the open literature, operating experience, professional contacts, and public participation. The GEIS examined the reasonably foreseeable range of environmental impacts that could occur as a result of renewing licenses of individual nuclear power plants. The GEIS, to the extent possible, established the bounds and significance of these potential impacts. The analyses in the GEIS encompassed all operating light-water power reactors. For each type of environmental impact, the GEIS established generic findings where possible. For some of the environmental impacts, a generic determination could not be made and those issues require a detailed site-specific analysis. While plant and site-specific information was used in developing the generic findings, the NRC did not intend the GEIS to be a compilation of individual plant environmental impact statements. The findings of the GEIS are codified in the NRC's environmental regulations in 10 CFR Part 51.

4.2.2.3   Why did the NRC pursue the development of a generic environmental impact statement (GEIS)?

The NRC's decision to develop the GEIS was based on three principal objectives:

- to provide an understanding of the types and severities of environmental impacts that may occur as a result of license renewal of nuclear power plants,

- to identify and assess those impacts that are expected to be generic to license renewal, and

- to support a rulemaking for 10 CFR Part 51 to define the number and scope of issues that need to be addressed by the applicants in plant-specific license renewal proceedings.

4.2.2.4 Why not consider each facility separately and fully review all issues at that facility?

Intake Trash Bars

The generic environmental impact statement (GEIS) was developed to establish an effective licensing process. It contains the results of a systematic evaluation of the environmental consequences of renewing an operating license and operating a nuclear power facility for an additional 20 years. Those environmental issues that could be resolved generically were analyzed in detail and were resolved in the GEIS. Those issues that were unique because of a site-specific attribute, a particular site setting or unique facility interface with the environment, or variability from site to site, were deferred and would be resolved at the time that an applicant sought license renewal. In the license renewal process, these issues are addressed by a site-specific supplement to the generic environmental impact statement (SEIS).

The GEIS is used to avoid duplication and allow the staff to focus specifically on those issues that are important for a particular plant (i.e., issues that are not generic). This is an appropriate and effective use of the concept of *tiering* that was promulgated by the President's Council on Environmental Quality (CEQ) in its 1978 regulations that implemented the requirements of National Environmental Policy Act (NEPA). (Tiering is discussed in the response to Question 4.1.6.)

A GEIS was developed for license renewal but not for initial construction and operation of the nuclear facilities because the licensing process for new reactors may involve consideration of land that would be disturbed, new demands placed on resources, and new discharges that require permits. However, for license renewal, there is a better understanding of the environmental equilibrium that has been established after a period of operation. In addition, in developing the GEIS, the NRC staff had the benefit of experience and information available for site-specific environmental impact statements that were performed for construction and operation of the plants.

The NRC's regulations require that a supplement to the Generic Environmental Impact Statement for License Renewal of Nuclear Plants be prepared for individual license renewal applications to address those impacts that could not be generically evaluated in the GEIS.

4.2.2.5 Why does the NRC develop a site-specific supplement to the generic environmental impact statement for license renewal (SEIS)?

The NRC's regulations require that a supplement to the *Generic Environmental Impact Statement for License Renewal of Nuclear Plants*, NUREG-1437 (GEIS) be prepared for individual license renewal applications to address those impacts that could not be generically evaluated in the GEIS. The GEIS addresses, in a generic fashion, the impacts associated with continued operation of a nuclear power facility for 20 years beyond its current license. The SEIS is a site-specific analysis of a facility that has requested a license renewal. The SEIS supplements the information provided in the GEIS.

4.2.3 The process for developing a site-specific supplement to the generic environmental impact statement for license renewal (SEIS)

4.2.3.1 How does the NRC conduct the environmental evaluation for license renewal?

The process for license renewal, as described in the NRC's regulations in 10 CFR Part 51, begins when an applicant submits the license renewal application containing an environmental report. After accepting the application, the NRC issues a Notice of Intent to prepare a site-specific supplement to the generic environmental impact statement for license renewal (SEIS) and conduct scoping. The Notice of Intent is posted on the NRC website and published in the *Federal Register*. The NRC also schedules a public scoping meeting in the vicinity of the facility (see response to Question 4.1.7 for a definition of scoping). Based on the scoping process and its independent review, the NRC staff issues a draft SEIS for public comment and holds a public meeting to discuss the findings in the draft SEIS and to obtain comments from the public and other interested stakeholders related to the draft report. The staff issues a final SEIS, which incorporates appropriate comments and changes. The final SEIS includes an appendix that presents the comments obtained at the public meeting and in writing and responds to those that are within the scope of the document.

> *The GEIS is used to avoid duplication and allow the staff to focus specifically on those issues that are important for a particular facility.*

4.2.3.2 What steps does the NRC have to take to complete the environmental review?

The environmental review for license renewal begins when an applicant sends its license renewal application to the NRC. One part of the application is the facility's environmental report. This is the starting point for the NRC staff's environmental review. The following activities occur after the license renewal application is received:

- The NRC staff places a notice in the *Federal Register* that the application has been received. The notice provides information to the public on how to access copies of the application. This notice is usually placed in the *Federal Register* within a month of the receipt of the application.

- The NRC staff places a second notice in the *Federal Register* approximately a month later. This notice indicates that the NRC staff has determined that the information in the application is sufficient and acceptable to begin the review. This determination is based on a comparison of the information provided in the application and the information required to be submitted by the regulations. Either this notice or a third notice placed in the *Federal Register* defines a minimum 60-day period for interested persons to file a request for a hearing or a petition to intervene.

- The NRC staff places a third (or fourth) notice in the *Federal Register* approximately 3 months after receiving the application. The purpose of this notice is to inform the public of the NRC's intent to prepare a site-specific supplement to the generic environmental impact statement on license renewal (SEIS) and to inform the public about the scoping process. The scoping process is conducted to define the proposed action, to determine the scope of the SEIS, and to identify the significant issues to be analyzed in depth. Public scoping meetings are held near the nuclear power reactor that is seeking license renewal. The *Federal Register* notice provides the times and locations of the public scoping meetings. The notice also provides addresses for written comments to be submitted in person, by mail, or electronically. The deadline for scoping comments is usually 60 days following the publication of the notice in the *Federal Register*.

The scoping process is a National Environmental Policy Act (NEPA) requirement conducted to define the proposed action, to determine the scope of the environmental review, and to identify the significant issues to be analyzed in depth.

- Approximately 120 days after the application is received, the NRC staff hosts two public scoping meetings on the same day within the vicinity of the nuclear power reactor being considered for relicensing. One meeting is in the afternoon and the other during the evening in an attempt to reach as many members of the public as possible. The meeting purpose, times, and locations are commonly advertised in local papers and on the radio to ensure that interested members of the public are aware of the public scoping meetings. Transcripts of the meetings are made available to the public approximately one month after the meeting is conducted.

- The NRC staff typically conducts a site audit at the facility and in the surrounding area at the time that it holds the scoping meetings. The purpose of this visit is to familiarize the NRC and contractor team with the site and its environs and to determine whether additional issues should be investigated as part of the license renewal environmental evaluation. The NRC/contractor team is composed of experts in the fields that are pertinent to the environmental review (see response to Question 4.2.3.7).

- The draft SEIS is published by the NRC after it completes its detailed reviews, approximately a year after the application is received. The NRC staff places a notice of availability in the *Federal Register* with instructions for the public and other interested parties on how to obtain copies. Copies are sent to all individuals on facility distribution lists. Individual copies are also provided to all members of the public upon request. The notice requests public comments on the draft supplement and provides addresses for delivering and sending the comments to the NRC. Usually, a 75-day period is provided for the public's review and the receipt of comments. The notice also alerts the public to a second set of public meetings to be held in the vicinity of the nuclear facility. The purpose of the meetings is to present an overview of the draft SEIS and to accept public comments on the document. Transcripts of the public meeting are made available approximately a month after the meeting is conducted.

Oyster Creek Nuclear Station Endangered Species Act Consultation with National Marine Fisheries Service

- Following receipt of the comments from the public, the applicant, and any interested local, state, Tribal, or Federal agencies, the NRC staff addresses the comments that are within scope of the draft SEIS and makes any appropriate changes. The final SEIS is published, including the list of the comments and the NRC staff's resolution for each comment. The final document is usually issued approximately 20 months after the application was received. The NRC staff then provides a final recommendation regarding the environmental review to the Commission or its designee.

4.2.3.3 Why does the NRC have a scoping process and what information is it specifically looking for during this process?

The scoping process is a National Environmental Policy Act (NEPA) requirement conducted to define the proposed action, to determine the scope of the environmental review, and to identify the significant issues to be analyzed in depth. Specifically, the scoping process for a site-specific supplement to the generic environmental impact statement on license renewal (SEIS) accomplishes the following:

- defines the proposed action,

- determines the scope of the SEIS and identifies the significant issues to be analyzed in depth,

- identifies and eliminates from detailed study those issues that are peripheral or that are not significant,

- identifies any environmental assessments and other environmental impact statements that are being or will be prepared that are related to but are not part of the scope of the SEIS,

- identifies other environmental review and consultation requirements related to the proposed action,

- indicates the relationship between the timing of the preparation of environmental analyses and the Commission's tentative planning and decision-making schedule,

- identifies any cooperating agencies and, as appropriate, allocates assignments for preparation and schedules for completing the SEIS, and

- describes how the SEIS will be prepared, including any contractor assistance to be used.

4.2.3.4 What are the technical areas included in the plant-specific review?

The NRC performs plant-specific reviews of the environmental impacts of license renewal in accordance with National Environmental Policy Act (NEPA) and the NRC's requirements in 10 CFR Part 51. The following technical areas are commonly included in the review:

Transmission Corridor

- land use,

- ground and surface water use,

- ground and surface water quality,

- air quality,

- aquatic resources,

- terrestrial resources,

- threatened and endangered species,

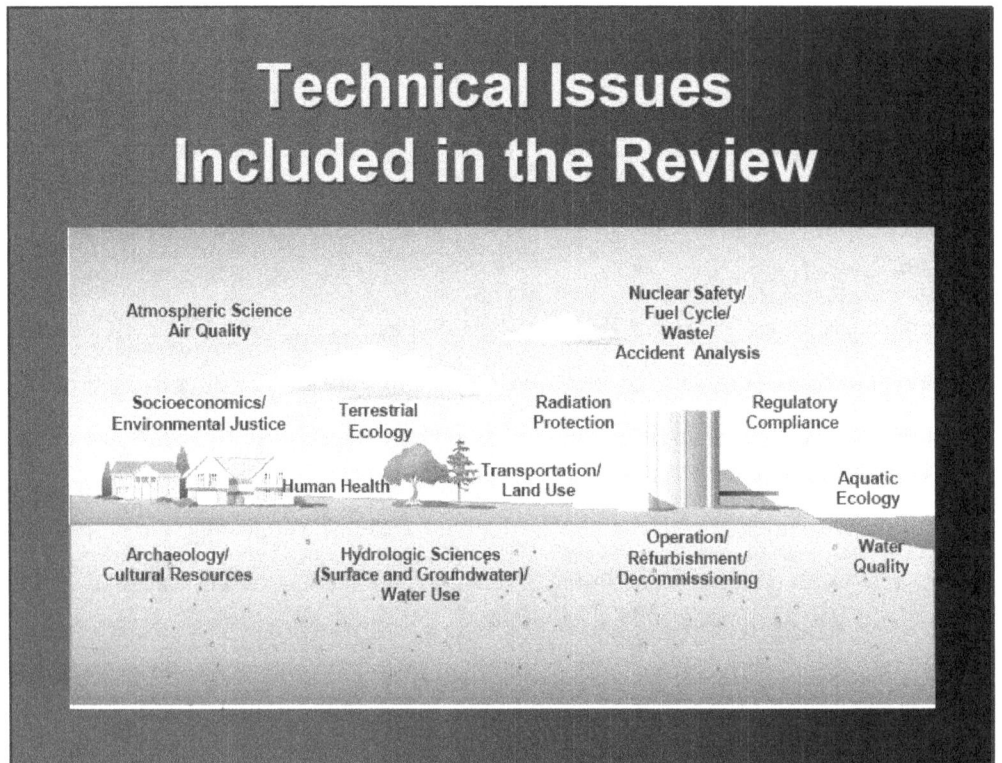

- radiological impacts.

- socioeconomic factors.

- environmental justice issues.

- historical and archaeological resources.

- related Federal project activities.

- postulated accidents.

- uranium fuel cycle and solid waste management.

- decommissioning.

- alternatives to license renewal, and

- irreversible or irretrievable resource commitments.

Other areas may be included as a result of information obtained during the NRC staff's review or from public comments during or following meetings that are held in the vicinity of the nuclear power reactor.

4.2.3.5   What is the geographical area that is considered in the review?

Environmental impacts are considered at the site itself and in surrounding areas that could be affected by the operation of the facility. Impacts include those along transmission systems that were built specifically to connect the facility to the grid. The facility's contribution to impacts from non-facility-related activities are also evaluated as cumulative impacts.

4.2.3.6 Are impacts considered only during the period of time encompassed by the renewed license?

The impacts are considered for refurbishment activities necessary for license renewal and for operational activities that take place during the period of the renewed license. Refurbishment activities are physical activities or changes to the facility or site that are undertaken to prepare a nuclear power facility for operation following license renewal. These activities, which occur as needed, include enhanced inspection, surveillance, testing, maintenance and repair, and replacement, modification, and refurbishment of plant systems, structures, and components. For some facilities, replacement of large components of the nuclear steam supply system (e.g., steam generator or pressurizer) may be necessary, as is repair or replacement of pumps, pipes, control rod systems, electronic circuitry, electrical and plumbing systems, or motors. Not many facilities are expected to need refurbishment activities in connection with license renewal. Many applicants anticipate that they will replace components and conduct additional inspection activities within the bounds of normal facility component replacement and inspection. None of the applications received to date (through the beginning of 2006) have identified any major facility refurbishment activities or modifications necessary to support the continued operation of the facility beyond the end of the existing operating license.

Steam Generator

4.2.3.7 What is the usual schedule for the environmental review of an application and who actually performs the environmental review?

The environmental review generally takes 20 months if no hearing has been granted. It is currently expected that the NRC staff will complete both the environmental and safety reviews and issue the renewed licenses within 30 months from receipt if a hearing is held or within 22 months from receipt if there is no hearing.

The environmental review is performed by a team of experts, including NRC staff members supported by contractor staff from national laboratories and other contractors. The team is composed of experts in a variety of fields, including:

- atmospheric science,
- hydrology (surface and groundwater use and quality),
- terrestrial ecology,
- aquatic ecology,
- land use,

The environmental review generally takes 20 months if no hearing has been granted.

- archaeology/cultural resources,
- socioeconomics/environmental justice,
- radiation protection,
- nuclear safety, and
- regulatory compliance.

Indian River Lagoon
St. Lucie Plant Site Audit

4.2.3.8   Which state, Tribal, county, or local agencies does the NRC contact during the review of the license applications?

State offices may be contacted for input during the NRC staff's analysis of the license application. Offices include organizations dealing with health and human services, cultural resources, and environmental protection and natural resources. The NRC staff may also contact county or local agencies, specifically those that may provide the staff with cultural and historic or socioeconomic information related to the staff's review of the license renewal application. The NRC staff also contacts Tribal nations that may have aboriginal ties to the land in the vicinity of the plant.

Although the NRC does not provide copies of the applicant's license renewal application to state, Tribal, county, or local agencies to review, the applicant may provide it directly to specific state offices. The NRC does post a notice indicating the receipt of the license renewal application in the *Federal Register* shortly after it receives the application. The notice indicates where copies are available and how they can be obtained. The NRC makes arrangements for a hard copy of the application to be available at a public library close to the site. Also, electronic copies are accessible from the NRC web site (see response to Question 5.2.7 for more information related to obtaining a copy of the application).

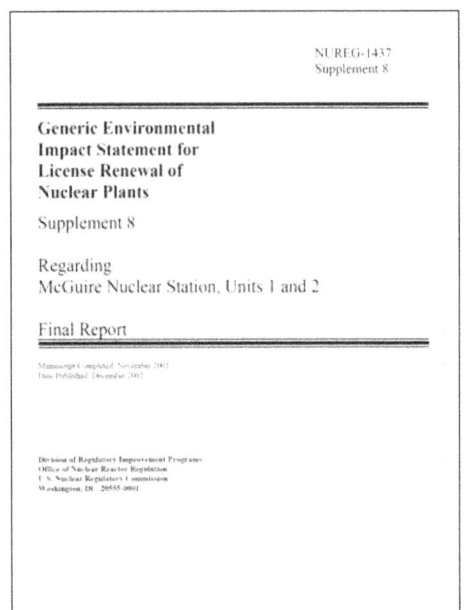

4.2.3.9   Where are the results of the environmental review published?

The results of the environmental review are published as a site-specific supplement to the generic environmental impact statement on license renewal (SEIS). The results are first published as a draft SEIS, which includes the NRC's analysis of the environmental impacts of the proposed license renewal action and the environmental impacts of the alternatives to the proposed action. The draft SEIS includes a preliminary recommendation regarding license renewal based on consideration of the information on the environmental impacts of license renewal and of alternatives contained in the SEIS. The staff then issues a final SEIS after considering public comments on the draft SEIS.

Instructions for obtaining an SEIS and the *Generic Environmental Impact Statement for License Renewal of Nuclear Plants* NUREG-1437 (GEIS) are given in the responses to Questions 5.2.8 and 5.2.9.

4.2.3.10 Who uses the site-specific supplement to the generic environmental impact statement on license renewal (SEIS), and how?

The SEIS is considered to be a disclosure document used to inform both decision-makers and the public of the environmental impacts of a specific action (in this case, license renewal). Once the NRC acts on the license renewal application, the NRC's SEIS becomes part of the licensing basis.

4.2.4 Categorization and Evaluation of Issues

4.2.4.1 What are Category 1 (generic) and Category 2 (site-specific) issues and why are they important in the analysis of environmental impacts?

The impact evaluation performed by the staff and presented in the *Generic Environmental Impact Statement for License Renewal of Nuclear Plants* NUREG-1437 (GEIS) identified 92 environmental issues that were considered for the license renewal evaluation for power reactors in the U.S. The industry, Federal, state, and local governmental agencies, members of the public, and citizen groups commented on and helped identify these 92 issues during the preparation of the GEIS. The 92 issues are listed in Appendix A of this document. For each of the identified 92 issues, the staff evaluated existing data from all operating power plants throughout the U.S. From this evaluation, the staff determined which issues could be considered generically and which issues do not lend themselves to generic consideration. The GEIS divides the 92 issues that were assessed into two principle categories: one for generic issues (which are termed "Category 1 issues") and the other for site-specific issues (termed "Category 2 issues"); two issues were not categorized.

Category 1 issues are termed "generic" issues because the conclusions related to their environmental impacts were found to be common to all plants (or, in some cases, to plants having specific characteristics). For Category 1 issues, a single level of significance was common to all plants, mitigation was considered, and the NRC determined that it was not likely to be beneficial.

Category 1 (generic) issues are those that meet all of the following criteria:

1) The environmental impacts associated with the issue have been determined to apply either to all plants or, for some issues, to plants having a specific type of cooling system or other specified plant or site characteristic.

2) A single significance level (i.e., SMALL, MODERATE, or LARGE) has been assigned to the impacts (except for collective offsite radiological impacts from the fuel cycle and from high-level waste and spent fuel disposal) for all plants. (See the response to Question 4.2.4.13 for a definition of the levels of significance.)

The Generic Environmental Impact Statement for License Renewal of Nuclear Plants NUREG 1437 (GEIS) identified 92 environmental issues that were considered for the license renewal evaluation for power reactors in the U.S.

3) Mitigation of adverse impacts associated with the issue has been considered in the analyses, and it has been determined that additional plant-specific mitigation measures are not likely to be sufficiently beneficial to warrant implementation.

Category 1 issues are termed "generic" issues because the conclusions related to their environmental impacts were found to be common to all plants (or, in some cases, to plants having specific characteristics such as a particular type of cooling system). For Category 1 issues, a single level of significance was common to all plants, mitigation was considered, and the NRC determined that it was not likely to be beneficial. Issues that were resolved generically are not reevaluated in the site-specific supplement to the generic environmental impact statement on license renewal (SEIS) because the conclusions reached would be the same as in the GEIS, unless new and significant information is identified that would lead the NRC staff to reevaluate the GEIS's conclusions. During the environmental reviews of license renewal applications, the NRC staff makes a concerted effort to determine whether any new and significant information exists that would change the generic conclusions for Category 1 issues.

Category 2 issues are those that require a site specific review.

Category 2 issues are those that require a site-specific review. For each of the Category 2 issues applicable to the site under review, the staff evaluates site-specific data provided by the applicant, other Federal agencies, state agencies, Tribal and local governments, as well as information from the open literature and members of the public. From this data, the staff makes a site-specific evaluation of the particular issues and presents its analyses and conclusions in the SEIS for the facility. Additionally, two uncategorized issues require site-specific considerations as well.

Natural Draft Cooling Tower

Three examples of Category 1 issues
- *Water quality impacts from biocides*
- *Bird collisions with cooling towers*
- *Aesthetic impacts*

4.2.4.2 What are some examples of generic (Category 1) issues and how were they determined to be generic?

The generic environmental impact statement for license renewal (GEIS) evaluated 92 environmental issues, and, of these, 69 were found to be generic (Category 1). Following are three examples of generic (Category 1) issues:

Example 1: The discharge of chlorine and other biocides is regulated by the National Pollutant Discharge Elimination System (NPDES) permit of each nuclear power facility. Regulatory concern about toxic effects of chlorine and its combination products, and the operating experience with control of biofouling organisms, such as mussels or clams, have led many facilities to eliminate the use of chlorine or reduce the amount used to below the levels that were originally anticipated in the environmental statements issued for construction or operation. Because of these refinements, water quality impacts from biocides was not a concern for regulatory and resource agencies provided that an applicant remained in compliance with the limits in the NPDES permit. Based on the literature, operational monitoring reports, consultations with licensees and regulatory agencies, and comments on the draft GEIS, water quality effects from the discharge of chlorine and other biocides are thus considered to be of small significance for all facilities. This issue was determined to be a generic (Category 1) issue.

Example 2: Bird collisions with cooling towers is of concern if the stability of the local population of any bird species is threatened or if the reduction of numbers within any bird population significantly impairs its function within the local ecosystem. Monitoring of bird collisions has occurred at several nuclear plants with natural draft cooling towers. The existing data suggest that cooling towers cause only a very small fraction of the total bird collision mortality in the U.S. and involve sufficiently small numbers for any species that it is unlikely that the losses would threaten the stability of local populations. There is also no reason to believe that the annual mortality rate resulting from collision of birds with any cooling tower would be different during the license renewal term. Thus, bird mortality was determined to be a generic (Category 1) issue.

> *Three examples of Category 2 issues*
> - *Entrainment of fish and shellfish at facilities with once through cooling systems*
> - *Threatened or endangered species*
> - *Historic and archaeological resources*

Example 3: The analysis of aesthetic impacts of license renewal involved staff examination of local perceptions at seven case study sites, a brief survey of the original and eventual aesthetic impacts at other operating nuclear power facilities, a survey of relevant academic literature, and a review of recent newspaper and magazine articles related to these issues. Nuclear power facilities, especially those with natural draft cooling towers, stand out starkly from their backgrounds, both physically and symbolically. The GEIS acknowledged that some people regard the existing facility structures and vapor plumes negatively. However, during license renewal, applicants are not expected to alter the existing visual intrusiveness of any facility. Thus, the extent of negative perceptions likely remains constant. Because negative views have not been sufficient to measurably impact community institutions and functions in the past, the staff concluded in the GEIS that the impacts on aesthetic resources would likely be small. Thus, this issue was determined to be generic (Category 1).

Green Turtle*

4.2.4.3 What are some examples of site-specific (Category 2) issues, and how were they determined to be site-specific?

Of the 92 environmental issues in the generic environmental impact statement for license renewal (GEIS), 23 issues need a site-specific review and analysis. Twenty-one of these are considered to be Category 2 issues. The remaining two issues, environmental justice and chronic effects of electromagnetic fields, were not categorized and are addressed by the site-specific analysis. Three examples of site-specific (Category 2) issues are given here.

Example 1: The entrainment of fish and shellfish in early life stages at facilities with once-through cooling systems requires a site-specific analysis for two reasons. First, a single significance level cannot be assigned. While the impacts may be small at many facilities, they may be moderate or even large at a few facilities with once-through cooling systems. Second, ongoing restoration efforts may increase the numbers of fish susceptible to intake effects during the license renewal period, so that entrainment studies conducted in support of the original license may no longer be valid. For these reasons, the entrainment of fish and shellfish is a site-specific (Category 2) issue for facilities with once-through cooling systems.

* Photo by Ursula Keuper-Bennett/Peter Bennett.

Example 2: Threatened or endangered species is a potentially relevant issue for all cooling system types and for transmission lines. The GEIS review showed that neither current cooling system operations nor electric power transmission lines associated with nuclear power facilities are having a significant adverse impact on any threatened or endangered species. However, widespread conversion of natural habitats and other human activities continues to cause the decline of native plants and animals. As biologists review the status of species, additional species threatened with extinction are being identified; consequently, it is not possible to ensure that future power facility operations will not be found to adversely affect some currently unrecognized threatened or endangered species. Future endangered species recovery efforts may require modifications of power facility operations. Without site-specific and project-specific information, the magnitude or significance of impacts on threatened and endangered species cannot be assessed or predicted. Thus, because no generic conclusion on the significance of potential impacts on endangered species can be reached, this is a site-specific (Category 2) issue.

Site specific issues (Category 2 issues) must be thoroughly analyzed.

Example 3: Historic and archaeological resources are considered to be a site-specific (Category 2) issue because site-specific information is needed to assess the significance of impacts to these resources. Determinations of impacts must be made through consultation with a State Historic Preservation Officer (SHPO). Any mitigation measures must likewise be determined on a case-by-case basis.

4.2.4.4 During the environmental review, how are generic issues and site-specific issues reviewed?

Generic and site-specific issues are dealt with differently during the environmental review. Issues that were resolved generically (Category 1) are not reevaluated in the site-specific supplement to the generic environmental impact statement on license renewal (SEIS) because the conclusions reached would be the same as in the generic environmental impact statement for license renewal (GEIS), unless new and significant information is identified that would lead the NRC staff to reevaluate the GEIS's conclusions. During the environmental review, the NRC staff makes a concerted effort to determine whether new and significant information exists for the specific site being evaluated that would change the generic conclusions for Category 1 issues.

Issues that were resolved generically (Category 1) are not reevaluated in the site specific supplement to the generic environmental impact statement on license renewal unless new and significant information is identified that would lead the NRC staff to reevaluate the GEIS's conclusions.

Site-specific issues (Category 2 issues) must be thoroughly analyzed by the applicant as part of its submittal and included in detail in its environmental report. The NRC staff then independently evaluates the issue as part of its SEIS.

4.2.4.5 Does the NRC take the Category 1 (generic) issues "off the table" for public review?

The NRC does not take the generic (Category 1) issues "off the table" for public review. If there is new and significant information that would change the conclusions reached in the generic environmental impact statement for license renewal (GEIS), then the staff notifies the Commission and the issue requires a site-specific analysis.

During the scoping process and the environmental review, the NRC is looking for any information that could demonstrate that there are unique characteristics related to the facility or the environment surrounding the facility that would lead to the conclusion that the generic determination for a particular issue is not valid for a specific site. The NRC staff discusses and evaluates potential new and significant information on impacts of operations during the renewal term in the site-specific supplement to the generic environmental impact statement on license renewal (SEIS).

New information can be identified from a number of sources, including the applicant, NRC review activities, other agencies, or public comments.

4.2.4.6   Are human health issues a "generic issue"? Why aren't they evaluated for each facility?

Not all human health issues are considered generic (Category 1). The radiological impacts on human health (both to the public and to plant workers) and noise are considered generic (Category 1) issues. However, the impacts of microbiological organisms on public health, under certain facility configurations, and the acute effects of electromagnetic fields are considered to be site-specific (Category 2) issues. The chronic effects of electromagnetic fields were not categorized as either generic or site-specific because research is continuing in this area and a consensus scientific view has not been reached.

For the generic issue of radiological impacts on human health, radiation doses to members of the public from the current operation of nuclear power facilities have been examined from a variety of perspectives, and the impacts were found to be well within design objectives and regulations in each instance. Because there is no reason to expect effluents to increase during the period of the renewal license, effluent levels during continued operation are expected to be well within regulatory limits. However, as with all Category 1 conclusions, the NRC staff review evaluates each application and the site to determine if there is new and significant information that would change the conclusion in the generic environmental impact statement for license renewal (GEIS). In addition, current mitigation practices have resulted in declining public radiation dose and are expected to continue to do so. The NRC staff concluded in the GEIS that the significance of radiation exposures to the public attributable to operation after license renewal will be small at all sites and that this is a generic (Category 1) issue.

Occupational doses attributable to normal operation during the license renewal term were also examined from several different perspectives. An estimate of a 5-8 percent increase in doses for the typical plant worker for the renewal period was made, based on the slight increase in radioactive inventories that occurs as a plant ages. Even with this increase, the anticipated doses will remain well below the regulatory limits. Therefore, occupational radiation exposure during the renewed license period meets the standard of small significance and thus would be a generic (Category 1) issue.

4.2.4.7   What issues are precluded from consideration?

A number of issues are not considered in the environmental review for license renewal conducted by the NRC, including but not limited to:

- safety,
- operational issues that require a separate National Environmental Policy Act (NEPA) review (such as an independent spent fuel storage installation),

- security and safeguard issues,

- emergency preparedness (including distribution of potassium iodide),

- need for power,

- spent fuel disposal and storage,

- economic feasibility, and

- cost-benefit analyses.

Section 4.3 discusses these issues in further depth and provides the reason why they are not considered in the environmental evaluation.

4.2.4.8   What if new information is revealed about an existing issue?

New information can be identified from a number of sources, including the applicant, NRC review activities, other agencies, or public comments. If new information is revealed about an existing issue, then the NRC evaluates the significance of that information by using regulatory guidance (such as NUREG-1555, Volumes 1 and 2) and calling upon experts from within the NRC, its contractors or other recognized institutions. If the new information concerns a generic issue (a Category 1 issue), then the NRC staff determines whether the new information indicates that the analysis supporting the NRC license renewal environmental protection rule is not correct, e.g., the impacts are beyond that described in the *Generic Environmental Impact Statement for License Renewal of Nuclear Plants*, NUREG-1437 (GEIS) and codified in the NRC's rule.

If the new and significant information is relevant to the particular plant and is also relevant to other plants, then the NRC staff will seek Commission approval to either suspend the application of the rule on a generic basis or delay granting the particular renewal application (and possibly other renewal applications) until the analysis in the GEIS is updated and the rule amended. If the rule is suspended, then each subsequent site-specific supplement to the generic environmental impact statement on license renewal (SEIS) will reflect the corrected analysis until such time as the rule is amended. If the new and significant information is relevant only to the particular plant, then the NRC staff will seek Commission approval to waive the application of the rule on a site-specific basis for that issue and perform an analysis to the level of detail that would be the equivalent of a site-specific issue (a Category 2 issue). The SEIS would reflect the corrected analysis, as appropriate.

4.2.4.9   What if a new issue is identified that has not been considered before?

New information can be identified from a number of sources, including the applicant, NRC review activities, other agencies, or public comments. If a new issue is revealed, then it is first analyzed to determine whether it is within the scope of the license renewal evaluation (see response to Question 4.2.4.7 for issues that the Commission has determined are not within the scope of license renewal). If a new environmental issue is determined to be within the scope of license renewal and it was not addressed in the *Generic Environmental Impact Statement for License Renewal of Nuclear Plants*, NUREG-1437 (GEIS) or codified in the NRC license renewal environmental protection rule, then the NRC evaluates the significance of the information by calling upon experts from within the NRC, its contractors or other recognized institutions. If the new issue is relevant only to a particular site, then the NRC staff will perform a site-specific analysis and include its conclusion in the site-specific supplement to the generic environmental impact statement on license renewal (SEIS). If the new and significant information appears to be relevant to other sites, then the NRC staff will consider the issue in future SEISs and include it as a candidate for evaluation in the periodic update of the GEIS and possible amendment to the rule.

4.2.4.10 What are some examples of new issues that have been discussed during license renewal?

Although a number of new issues have been raised and evaluated by the NRC and the results of the evaluation documented in the appropriate site-specific supplements to the generic environmental impact statement on license renewal (SEIS), the issues that were identified were not significant and did not invalidate any of the staff's generic determinations. Two examples follow.

Example 1: During the scoping meetings on the Calvert Cliffs license application, a member of the public raised the issue of extremophiles, which are microbiological organisms that live in high-radiation and high-temperature environments. The NRC staff evaluated this issue and determined that, while it was new information, it was not significant because extremophiles would not likely be able to survive and compete with the indigenous microbiota of the relatively cold waters of Chesapeake Bay, once cooling water was discharged from the Calvert Cliffs facility.

Example 2: A new issue was identified by the staff during its review of the North Anna license renewal application. The staff identified a potential issue related to the nuisance species water hyacinth (*Hydrilla verticillata*), a submerged, aquatic macrophyte (large plant) that inhabits many freshwater rivers, lakes, and ponds in North America. Although higher water temperatures can increase the growing season of water hyacinths, the staff concluded that the issue was not significant because grass carp appeared to be effectively controlling the growth and biomass of the water hyacinth.

Hydrilla verticillata

4.2.4.11 Who decides if a comment contains new and significant items that have been discussed during license renewal?

The NRC staff determines whether an issue is new and significant after a thorough evaluation of the issue. New and significant information would involve an environmental issue that was not covered in the generic environmental impact statement for license renewal (GEIS) or codified in NRC's regulations and that could materially affect the Commission's earlier conclusion. It could also be information that was considered in the analyses summarized in the GEIS but that leads to a finding of environmental impact that is different from the finding presented in the GEIS (and codified in 10 CFR Part 51).

The NRC staff determines whether an issue is new and significant after a thorough evaluation of the issue.

4.2.4.12 What are some examples of new and significant items that have been discussed during license renewal?

As of the beginning of 2006, there have not been any examples of new issues that were also considered significant. This does not mean that there are no "new and significant" issues that will be found during the license renewal evaluation process. However, it is an indication that the review and analyses performed in preparing the generic environmental impact statement for license renewal (GEIS) were comprehensive and thorough. Both the public scoping sessions and the site evaluations are expected to remain essential parts of the license renewal process for raising issues for consideration.

4.2.4.13 What is meant by significance and what do the three levels of significance (SMALL, MODERATE, and LARGE) mean?

Significance indicates the importance of likely environmental impacts. The determination of significance is made by considering two variables: context and intensity. Context is the geographic, biophysical, and social context in which the effects will occur. In the case of license renewal, the context is the environment surrounding the facility. Intensity refers to the severity of the impact, in whatever context it occurs.

> ***Significance*** *indicates the importance of likely environmental impacts and is determined by considering two variables: context and intensity.*
>
> ***Context*** *is the geographic, biophysical, and social context in which the effects will occur.*
>
> ***Intensity*** *refers to the severity of the impact, in whatever context it occurs.*

The NRC developed a three-level standard of significance – SMALL, MODERATE, and LARGE – using the President's Council on Environmental Quality (CEQ) guidelines:

SMALL – environmental impacts are not detectable or are so minor than they will neither destabilize nor noticeably alter any important attribute of the resource.

MODERATE – environmental impacts are sufficient to alter noticeably but not destabilize important attributes of the resource.

LARGE – environmental effects are clearly noticeable and are sufficient to destabilize important attributes of the resource.

4.2.4.14 Who decides if an impact is of SMALL, MODERATE, or LARGE significance?

For Category 1 issues, the NRC staff assigned a significance level to each environmental issue analyzed in the generic environmental impact statement for license renewal (GEIS). The discussion of each environmental issue in the GEIS includes an explanation of how the significance category was determined. The determination of the significance category was made independently of the consideration of the potential benefit of additional mitigation.

For the Category 2 issues, the uncategorized issues, and the newly identified issues, the NRC will assign the significance level after an in-depth evaluation.

4.2.4.15 What are cumulative impacts, and how does NRC evaluate cumulative impacts?

Cumulative impacts on the environment result when impacts of an action are added to other past, present, and reasonably foreseeable future actions. Cumulative impacts can result from individually small impacts that become significant when taken collectively over a geographic area or a period of time. Any agency (Federal or non-Federal) or non-governmental activities can contribute through their actions or approvals to cumulative effects. These combined impacts are defined as "cumulative" and include individually minor but collectively significant actions taking place over a geographic area or a period of time.

Cumulative impacts on the environment result when impacts of an action are added to other past, present, and reasonably foreseeable future actions.

The NRC evaluates cumulative effects during the site visit and scoping process by identifying the impacts that have affected the environment surrounding the facility. For example, the close proximity of another nuclear reactor facility or another industrial facility that also discharges warm water into the same river may have a cumulative impact on aquatic ecology that is greater than the impact of just one facility. The staff would take into consideration the potential for cumulative impacts from both facilities.

4.2.4.16 How does the NRC consider environmental justice in its environmental review?

On February 11, 1994, the President issued Executive Order 12898, *Federal Actions to Address Environmental Justice in Minority Populations and Low-Income Populations*. This Order requires each Federal Executive Branch to identify and address, as appropriate, disproportionately high and adverse human health or environmental effects on minority and low-income populations resulting from its actions. The memorandum accompanying the Executive Order directed Federal executive agencies to consider environmental justice. The President's Council on Environmental Quality (CEQ) provided guidance for addressing environmental justice. Although the Executive Order is not mandatory for independent agencies, the NRC has voluntarily committed to undertake environmental justice reviews as part of its National Environmental Policy Act (NEPA) responsibilities. The Commission's "Policy Statement on the Treatment of Environmental Justice Matters in NRC Regulatory and Licensing Actions" contains guidance and information for addressing issues of environmental justice (69 FR 52040). Specific guidance was formulated by the NRC staff and is found in NRC Office of Nuclear Reactor Regulation Office Instruction LIC-203, *Procedural Guidance for Preparing Environmental Assessments and Considering Environmental Issues.*

Addendum 1 of the GEIS documented the NRC staff's analysis of the potential cumulative impacts of transporting spent nuclear fuel in the vicinity of a single high level waste repository.

In order to perform a review of environmental justice in the vicinity of a nuclear power plant, the NRC staff examines the geographic distribution of minority and low-income populations within 80 kilometers (50 miles) of the site. The staff uses the most recent census data available. The staff also supplements its analysis by field inquiries to such groups as county planning departments, social service agencies, agricultural extension personnel, and private social service agencies. Once the locations of minority and low-income populations are identified, the staff evaluates whether any of the environmental impacts of the proposed action could affect these populations in a disproportionately high and adverse manner.

Mitigation is:
- *avoiding,*
- *minimizing,*
- *rectifying or*
- *eliminating an impact or*
- *compensating for the impact*

4.2.4.17 What is mitigation, and who decides if it is necessary?

According to the President's Council on Environmental Quality (CEQ) (40 CFR 1508.20), mitigation means:

- avoiding the impact altogether by not taking a certain action or parts of an action,

- minimizing impacts by limiting the degree or magnitude of the action and its implementation,

- rectifying the impact by repairing, rehabilitating, or restoring the affected environment,

- reducing or eliminating the impact over time by preservation and maintenance operations during the life of the action, and

- compensating for the impact by replacing or providing substitute resources or environments.

In terms of the impacts during license renewal, this definition can include such activities as:

- using best-management practices to mitigate the impact of any required dredging,

- relocating a project, such as additional storage or lay-down yards, to avoid impact on a historic or an archaeological site,

- reconfiguring intake structures to reduce impingement or entrainment of fish and shellfish larvae, and

- making structural changes to equipment to mitigate the potential for severe accidents.

4.2.4.18 Could the generic environmental impact statement for license renewal (GEIS) become outdated and is it ever updated?

The environmental review does take into account the environmental effects of postulated plant accidents that might occur during the license renewal term.

In 10 CFR Part 51, the Commission anticipated the need to revisit the GEIS and its implementing regulations. The Commission declared its intent to revisit the GEIS on a 10-year cycle to determine whether the technical bases or conclusions needed to be updated. The GEIS represents a snapshot in time. Therefore, it is appropriate to periodically determine whether changes have occurred that should be included in an update to the GEIS. Science and the natural environment evolve and the scientific community's understanding of issues, methods, and assumptions may need to be revisited. Experience gained in using the regulatory framework may identify situations in which the NRC staff has used less than optimal approaches to address issues and to state conclusions. Changes in statutes, regulations, policies, and practices and the structure of the power market may have a cascading impact on the NRC licensing framework.

An example of the past updating of the GEIS is the addendum that was published in 1999. This addendum documented the NRC staff's analysis of the potential cumulative impacts of transporting spent nuclear fuel in the vicinity of a single high-level waste repository. It summarized the staff's analyses of the environmental impacts of the transportation of higher enrichment and higher burn-up spent nuclear fuel, which were found to be consistent with the values in Table S–4 of 10 CFR 51.52.

Currently, the GEIS for license renewal, which was originally issued in 1996 as NUREG-1437, is being updated. A notice of intent to update the GEIS was published in the *Federal Register* on June 3, 2003 (68 FR 33209). Public scoping meetings occurred during July 2003 in four regional locations (Atlanta, Georgia; Oak Lawn, Illinois; Anaheim, California; and Boston, Massachusetts), and the comments obtained from these meetings and from letters and emails sent to the NRC are being evaluated to determine whether and how the GEIS should be updated. The scoping process helps the NRC staff identify and eliminate from a detailed study those issues that are peripheral, that are not in scope, or that have been covered by other environmental reviews. The NRC is considering these comments and will factor the appropriate issues into a draft updated GEIS and into proposed rule changes if necessary. After all interested stakeholders (including the public) have had an opportunity to comment on the draft of the GEIS and the proposed rule, the NRC will issue a final GEIS and a final rule.

4.2.4.19 What kind of changes might be made in the generic environmental impact statement for license renewal (GEIS)?

As a framework, the NRC staff compiled a list of issues that may prompt changes in the GEIS, including:

- new and significant information (see responses to Questions 4.2.4.11 and 4.2.4.12),

- changes in NRC staff practices resulting from legislative or industry actions, for example, the designation of Yucca Mountain as the repository for spent nuclear waste in Public Law 107-200, 116 Stat. 735 (2002) (see response to Question 4.5.2),

- statutory or regulatory changes, for example, the U.S. Environmental Protection Agency's (EPA's) regulations establishing performance standards on cooling water intake structures for existing facilities,

- industry structural changes, for example, changes in the regulation of the power market or the distinctions between generators and distributors of power, which may have some bearing on the influence or control over activities that the current license holder may have as compared with that of the original license holder,

- incorrect characterizations that occurred in the GEIS, for example, the statement that license renewal is a major Federal action significantly affecting the quality of the human environment (see response to Question 4.2.1.2),

- omitted issues, for example, the impacts associated with dredging activities that may occur periodically and within the period of extended operation,

- confusion, for example, confusion between the impacts from severe accidents, which is a generic (Category 1) issue, and the analysis of severe accident mitigation alternatives, which is a site-specific (Category 2) issue, and

- realignment to improve clarity, for example, of the 92 specific issues listed in the GEIS, some are listed twice – once pertaining to the renewal process and again pertaining to refurbishment. Other issues are listed once, with a statement that they apply to both refurbishment and renewal. This can be confusing for a reader who is trying to understand how many issues are involved.

4.2.4.20 How can you process license renewal applications when you are updating the generic environmental impact statement for license renewal (GEIS)?

New applications will be evaluated under the existing regulatory framework using the GEIS as previously published and codified in NRC's regulations. However, insights and information gained during the GEIS update process and from experience with the completed license renewals using the GEIS will be used during the review of ongoing and upcoming applications until the update of the GEIS and appropriate revisions to 10 CFR Part 51 are completed.

The scope consists of the range of actions, alternatives, and impacts to be considered in an environmental impact statement.

4.3 Scope of the Environmental Review for License Renewal

Some of the most frequent questions from members of the public relate to what issues are considered within the scope of license renewal. Defining the scope is an important step because it allows the NRC to concentrate on the essential issues of the action being considered. This section answers some of the most common questions related to the scope of the environmental review for license renewal.

4.3.1 Why are there limits on the scope of the environmental review?

The scope consists of the range of actions, alternatives, and impacts to be considered in an environmental impact statement. The purpose of scoping is to identify the significant issues related to a proposed action. Scoping also identifies and eliminates from detailed study those issues that are not significant or have

The National Environmental Policy Act process focuses on environmental impacts rather than on issues related to the safety of an operation. Safety issues become important to the environmental review when they could result in environmental impacts. As a result, the environmental effects of postulated accidents are considered in the site specific supplement.

been covered by a prior environmental review. Having a defined scope for the environmental review allows the NRC to concentrate on the essential issues of actions being considered rather than on issues that may have been or are being evaluated in different regulatory review processes, such as a safety review.

4.3.2 Why are safety issues outside the scope of the environmental review?

The National Environmental Policy Act (NEPA) process focuses on environmental impacts rather than on issues related to the safety of an operation. Safety issues become important to the environmental review when they could result in environmental impacts, which is why the environmental effects of postulated accidents are considered in the site-specific supplement to the generic environmental impact statement on license renewal (SEIS). Because the NEPA regulations do not include a safety review, the NRC has codified the regulations for conducting an environmental impact statement separate from the regulations for reviewing safety issues during license renewal. The regulations governing the environmental review are in 10 CFR Part 51 and the regulations covering the safety review are in 10 CFR Part 54. For this reason, the license renewal process includes an environmental review that is distinct and separate from the safety review. Because the two reviews are separate, operational safety issues and safety issues related to aging are considered outside the scope for the environmental review, just as the environmental issues are not considered as part of the safety review. However, safety issues that are raised during the environmental review are forwarded to the appropriate NRC organization for consideration and appropriate action.

4.3.3 Accidents can cause environmental impacts, so does the environmental review consider accidents?

The environmental review does take into account the environmental effects of postulated plant accidents that might occur during the license renewal term. It also includes a review of the alternatives to mitigate severe accidents if this has not previously been evaluated for the applicant's plant (see Section 4.4). The purpose of this consideration is to ensure that plant changes (i.e., hardware, procedures, and training) with the potential for improving severe accident safety performance are identified, evaluated, and, if appropriate, implemented. As a result, the impacts of accidents are considered within the scope of the environmental review for license renewal.

4.3.4 Why are security issues outside the scope of the environmental review?

Security issues need to be dealt with constantly as a part of the current (and renewed) operating license.

Security issues such as safeguards planning are not tied to a license renewal action but are considered to be issues that need to be dealt with constantly as a part of the current (and renewed) operating license. Security issues are periodically reviewed and updated at every operating plant. These reviews continue throughout the period of an operating license, whether the original or renewed license. If issues related to security are discovered at a nuclear plant, they are addressed immediately, and any necessary changes reviewed and incorporated under the operating license.

4.3.5 Why are acts of terrorism considered outside the scope of the environmental review?

The NRC and other Federal agencies have heightened vigilance and implemented initiatives to evaluate and respond to possible threats posed by terrorists, including the use of aircraft against commercial nuclear power facilities and independent spent fuel storage installations (as discussed in the response to Question 4.8.1). Malevolent acts remain speculative and beyond the scope of a National Environmental Policy Act (NEPA) review. The NRC routinely assesses threats and other information provided by other Federal agencies and sources. The NRC also ensures that licensees meet appropriate security-level requirements. The NRC will continue to focus on prevention of terrorist acts for all nuclear facilities and will not focus on site-specific evaluations of speculative environmental impacts resulting from terrorist acts. While these are legitimate matters of concern, they will continue to be addressed through the ongoing regulatory process as a current and generic regulatory issue that affects all nuclear facilities and many of the activities conducted at nuclear facilities. The issue of security and risk from malevolent acts at nuclear power facilities is not unique to facilities that have requested a renewal to their licenses.

The NRC will continue to focus on prevention of terrorist acts through the ongoing regulatory process as a current and generic regulatory issue that affects all nuclear facilities.

4.3.6 Why are emergency preparedness questions outside the scope of the environmental review?

The Commission considered the need for a review of emergency planning issues in the context of license renewal during its rulemaking proceedings on 10 CFR Part 54, which included public notice and comment. As discussed in the Statement of Consideration for rulemaking (56 FR 64966), the programs for emergency preparedness at nuclear power facilities apply to all nuclear power facility licensees and require the specified levels of protection from each licensee regardless of plant design, construction, or license date. Requirements related to emergency planning are in the regulations at 10 CFR 50.47 and Appendix E to 10 CFR Part 50. These requirements apply to all operating licenses and will continue to apply to facilities with renewed licenses. Through its standards and required exercises, the Commission reviews existing emergency preparedness plans throughout the life of any facility, keeping up with changing demographics and other site-related factors. Therefore, the Commission has determined that there is no need for a special review of emergency planning issues in the context of an environmental review for license renewal.

4.3.7 Why is need for power outside the scope of the environmental review?

The regulatory authority over licensee economics (including the need for power) falls within the jurisdiction of the states and to some extent within the jurisdiction of the Federal Energy Regulatory Commission. The proposed rule for license renewal had included a cost-benefit analysis and consideration of licensee economics as part of the National Environmental Policy Act (NEPA) review. However, during the comment period, state, Federal, and licensee representatives expressed concern about the use of economic costs and cost-benefit balancing in the proposed rule and the generic environmental impact statement for license renewal (GEIS). They noted that President's Council on Environmental Quality (CEQ) regulations interpret NEPA to require only an assessment of the cumulative effects of a proposed Federal action on the natural and man-made environment and that the determination of the need for generating capacity has always been the states' responsibility. For this reason, the purpose and need for the proposed action (i.e., license renewal) is defined in the GEIS as follows:

> The purpose and need for the proposed action (renewal of an operating license) is to provide an option that allows for power generation capability beyond the term of a current nuclear power plant operating license to meet future system generating needs, as such needs may be determined by State, licensee, and, where authorized, Federal (other than NRC) decision-makers.

10 CFR 51.95(c)(2) states that

> the supplemental environmental impact statement for license renewal is not required to include discussion of need for power or the economic costs and economic benefits of the proposed action except insofar as such benefits and costs are either essential for a determination regarding the inclusion of an alternative in the range of alternatives considered or relevant to mitigation.

4.3.8 How do I get answers to my questions that fall outside the scope of the environmental review from the NRC?

There are three different ways for members of the public to receive answers to questions that fall outside the scope of the environmental review:

Public Meeting

- Public meetings – Members of the public are invited to plant-specific public meetings (see response to Question 5.1.3), where NRC staff members are available to answer any questions related to NRC-regulated activities that members of the public may have, including those that are outside the scope of the environmental review.

- NRC website – Answers to many questions that are outside the scope of the environmental review are also found on the NRC website, www.nrc.gov. The NRC has also developed a number of Frequently Asked Questions documents as well as informational brochures and fact-sheets to address issues that are of concern to the public. These documents, brochures, and fact sheets are also located on the NRC website.

- NRC environmental project manager – For plant-specific questions that are outside the scope, members of the public could contact the environmental project manager assigned by the NRC for the license renewal review for the specific plant. The phone number for each of the NRC environmental project managers is given on the NRC website, as well as in *Federal Register* notices and at the public meetings. The NRC environmental project manager can either answer questions or direct callers to the appropriate person in the agency for responding to their questions that are outside the scope of the review.

4.4 Severe Accident Mitigation Alternatives (SAMAs) Review

An analysis of SAMAs is included as part of the environmental review of the application for license renewal if it had not been considered earlier for the facility. A definition of SAMAs and an explanation of why they are included in the environmental review are included in this section. In addition, the process used to evaluate SAMAs and the types of changes that may occur in the plant as a result of the analysis are also discussed.

Severe accidents are those that could result in substantial damage to the reactor core, whether or not there are serious off-site consequences.

4.4.1 What is a Severe Accident Mitigation Alternatives (SAMAs) review?

The SAMAs review is an evaluation of alternatives to mitigate severe accidents. Severe accidents are those that could result in substantial damage to the reactor core, whether or not there are serious off-site consequences. The NRC staff reviews and evaluates SAMAs to ensure that changes that could improve severe accident safety performance are identified and evaluated. Potential improvements could include hardware modifications, changes to procedures, and changes to the training program.

A Severe Accident Mitigation Alternatives review is an evaluation of alternatives to mitigate severe accidents.

In some cases, SAMAs may have already been evaluated by the NRC staff in a previous environmental impact statement (EIS), supplement, or environmental assessment (EA) written for a facility before the applicant applied for license renewal. In such cases, the evaluation does not have to be repeated for that particular facility, according to NRC regulations in 10 CFR 51.53. However, if the NRC staff has not previously evaluated SAMAs for an applicant's plant in an EIS, a supplement, or an EA, the license renewal applicant is required to consider alternatives to mitigate severe accidents as part of the license renewal application.

4.4.2   Why are SAMAs considered part of the environmental review?

The Commission's regulations (10 CFR Part 51) require that license renewal applicants consider alternatives to mitigate severe accidents if the NRC staff has not previously evaluated severe accident mitigation alternatives (SAMAs) for the applicant's plant in an environmental impact statement (EIS) or related supplement or in an environmental assessment. This requirement results, in part, from a court decision partially reversing a Commission decision upholding an Atomic Safety and Licensing Board (ASLB) determination that it need not adjudicate a contention alleging that the NRC staff must consider, under the National Environmental Policy Act (NEPA), severe accident mitigation design alternatives in an environmental impact statement for an operating license (Limerick Ecology Action v. NRC, 869 F.2d 719 [3d Cir. 1989]). At the time that this requirement was established, licensees were identifying individual plant vulnerabilities and considering cost-beneficial improvements, e.g., in the individual plant examination (IPE) and the individual plant examination for external events (IPEEE) programs. The Commission believes that it is unlikely that any site-specific consideration of severe accident mitigation alternatives for license renewal would identify major plant design changes or modifications proving to be cost-beneficial for reducing severe accident frequency or consequences. However, because these programs had not yet been completed at all plants, the Commission considered a general conclusion regarding severe accident mitigation to be premature, and required a site-specific consideration of SAMAs for license renewal for plants lacking previously evaluated SAMAs in an EIS, related supplement, or environmental assessment.

4.4.3   What is the process for the Severe Accident Mitigation Alternatives (SAMAs) review?

The evaluation of SAMAs is a four-step process, as shown in Figure 4.1. The first step is to characterize overall plant risk and the leading contributors to the risk. This typically involves the extensive use of a plant-specific probabilistic safety assessment (PSA) study. The PSA identifies the different contributors of system failures and human errors that would be required for an accident to progress either to core damage or to containment failure.

The second step is to identify potential improvements that could reduce the risk. Information from the PSA, such as dominant accident sequences, equipment failures, and operator actions is used to identify plant improvements that would have the greatest impact in reducing risk. Improvements identified in other NRC and industry studies, as well as SAMAs analysis for other plants, are also considered in this process.

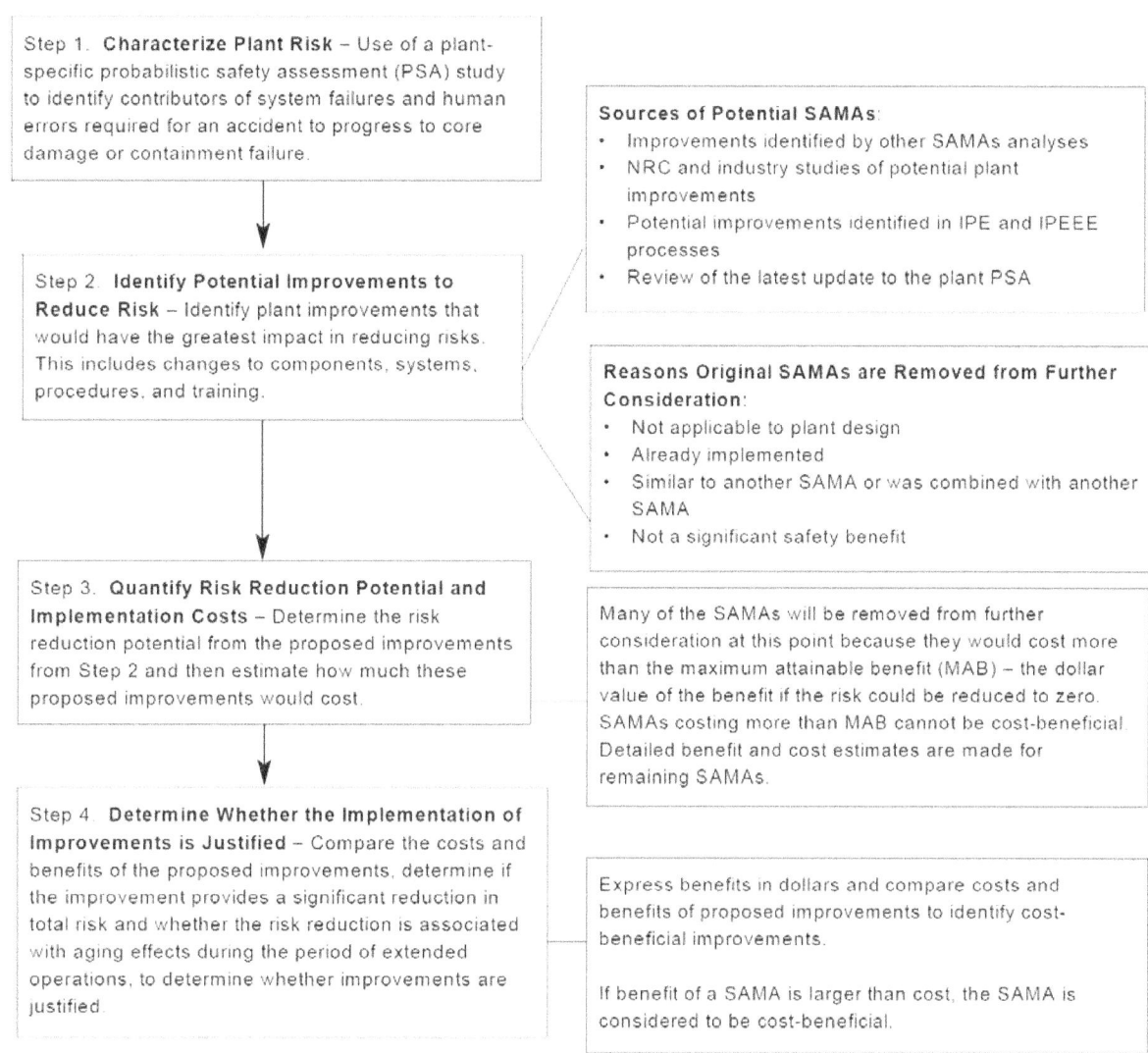

Step 1. Characterize Plant Risk – Use of a plant-specific probabilistic safety assessment (PSA) study to identify contributors of system failures and human errors required for an accident to progress to core damage or containment failure.

Step 2. Identify Potential Improvements to Reduce Risk – Identify plant improvements that would have the greatest impact in reducing risks. This includes changes to components, systems, procedures, and training.

Step 3. Quantify Risk Reduction Potential and Implementation Costs – Determine the risk reduction potential from the proposed improvements from Step 2 and then estimate how much these proposed improvements would cost.

Step 4. Determine Whether the Implementation of Improvements is Justified – Compare the costs and benefits of the proposed improvements, determine if the improvement provides a significant reduction in total risk and whether the risk reduction is associated with aging effects during the period of extended operations, to determine whether improvements are justified

Sources of Potential SAMAs:
- Improvements identified by other SAMAs analyses
- NRC and industry studies of potential plant improvements
- Potential improvements identified in IPE and IPEEE processes
- Review of the latest update to the plant PSA

Reasons Original SAMAs are Removed from Further Consideration:
- Not applicable to plant design
- Already implemented
- Similar to another SAMA or was combined with another SAMA
- Not a significant safety benefit

Many of the SAMAs will be removed from further consideration at this point because they would cost more than the maximum attainable benefit (MAB) – the dollar value of the benefit if the risk could be reduced to zero. SAMAs costing more than MAB cannot be cost-beneficial. Detailed benefit and cost estimates are made for remaining SAMAs.

Express benefits in dollars and compare costs and benefits of proposed improvements to identify cost-beneficial improvements.

If benefit of a SAMA is larger than cost, the SAMA is considered to be cost-beneficial.

Figure 4.1. Generalized Process for Identifying and Evaluating Potential Severe Accident Mitigation Alternatives (SAMAs)

The third step is to quantify the risk-reduction potential and the implementation cost for each of the improvements. The risk reduction is typically estimated using a conservative analysis that generally overestimates the risk-reduction potential by assuming that the plant improvement is highly effective in eliminating the accident sequence that the improvement is intended to address. Implementation costs are generally underestimated by neglecting certain cost factors, such as maintenance costs or surveillance costs associated with the plant modification. Overestimating the risk-reduction potential and under estimating the implementation costs in this step make it more likely that a potentially useful safety improvement would be retained for further consideration in the final step.

The risk-reduction potentials and the implementation cost estimates are used in the final step, which is to determine whether implementation of any of the improvements is justified. In determining whether the improvement is justified, the NRC staff looks at three factors: whether the improvement is cost-beneficial, in other words, whether the estimated benefit is greater than the estimate of the implementation cost of the SAMAs; whether the improvement provides a significant reduction in total risk, in other words, whether it eliminates a sequence or containment failure mode that contributes to a large fraction of plant risk; and whether the risk reduction is associated with aging effects during the period of extended operation, which would be an improvement implemented as part of the license renewal process.

4.4.4 What is the outcome of the review?

The Severe Accident Mitigation Alternatives analysis results in a list of plant improvements that meet the criteria of:

- *being cost beneficial*
- *providing a significant reduction in total risk, and*
- *being associated with aging effects during the period of extended operation.*

The outcome of the Severe Accident Mitigation Alternatives (SAMAs) analysis is a list of plant improvements that meet the criteria of being cost-beneficial, provide a significant reduction in total risk, and are associated with aging effects during the period of extended operation.

In some cases, however, the review leads to a determination that there are no specific SAMA candidates that are cost-beneficial. This may be the case where there is a low residual level of risk and where the applicant has, in fact, already implemented many plant improvements. In other cases, a SAMA that is potentially cost-beneficial may not relate to adequately managing the effects of aging during the period of extended operation. Such SAMAs need not be implemented as part of the license renewal pursuant to 10 CFR Part 54.

4.4.5 Does the applicant have to implement any identified changes?

The SAMAs analyses that have been performed to date have found SAMAs that were cost beneficial, or at least possibly cost beneficial subject to further analysis, in approximately half of the plants. None of the SAMAs identified related to managing the effects of aging during the period of extended operation.

The only changes that must be implemented by the applicant as part of the license renewal process are those that are identified as being cost-beneficial, that provide a significant reduction in total risk, *and* that are related to adequately managing the effects of aging during the period of extended operation. However, the Severe Accident Mitigation Alternatives (SAMAs) evaluation may identify some plant enhancements that appear to be cost-beneficial but that are not related to adequately managing the effects of aging during the period of extended operation. Such enhancements are considered as current operating issues and are further evaluated as changes that might appropriately be made under the current operating license rather than as a license renewal issue.

4.4.6 Have any changes been implemented at a plant as a result of the Severe Accident Mitigation Alternatives (SAMAs) review?

The SAMAs analyses that have been performed to date have found SAMAs that were cost-beneficial, or at least possibly cost-beneficial subject to further analysis, in approximately half of the plants. However, none of the SAMAs identified related to managing the effects of aging during the period of extended operation. Therefore, they did not need to be implemented as part of license renewal, pursuant to the regulations in 10 CFR Part 54. In general, the cost-beneficial SAMAs were identified for further evaluation by the licensee under the current operating license. In several cases, the applicant has decided to implement the modifications even though they were not related to license renewal.

4.5   Storage and Disposal of Spent Nuclear Fuel

Although the storage and disposal of high-level waste are not within the scope of environmental issues pertaining to license renewal, questions about these topics are asked frequently during public meetings and other opportunities for public comment. In the interest of providing a full picture of the issues associated with nuclear power facilities, this section provides information about the Nuclear Waste Policy Act and the status of Yucca Mountain as a repository for spent nuclear fuel from commercial reactors and the storage of spent fuel at nuclear power facilities.

4.5.1   What is the Nuclear Waste Policy Act?

The Nuclear Waste Policy Act establishes the Federal government's responsibility to provide a place for the permanent disposal of high-level radioactive waste and spent nuclear fuel and the generators' (commercial nuclear power facilities') responsibility to bear the costs of permanent disposal. The Act authorizes and requires the U.S. Department of Energy (DOE) to locate and build a permanent repository and an interim storage facility and develop a transportation system to safely link nuclear plants to the repository and interim storage facility. The Act was signed into law by President Reagan on January 7, 1983. The Act obligated DOE to begin disposal of high-level radioactive waste from commercial nuclear facilities by January 31, 1998. To date, an application for licensing these facilities has not been submitted to the NRC.

4.5.2   What is the status of Yucca Mountain?

For over two decades, research has been conducted to determine whether a site near Yucca Mountain, Nevada, is suitable for safely isolating highly radioactive nuclear waste. Four major steps – site characterization, site approval, licensing review, and construction – have to be completed before operation of the proposed high-level waste repository could occur. The first step, site characterization, including excavation of exploratory tunnels and testing of groundwater, has been completed. The result of the characterization study was documented by the development of an environmental impact statement (EIS). A draft EIS was

Yucca Mountain, Nevada

published in 1999 and a supplemental EIS in 2001. The final EIS was published along with a site recommendation in 2002.

The second step is site approval. Following publication of the final EIS and the U.S. Department of Energy's (DOE's) determination that the site is scientifically suitable for a geologic repository, President Bush recommended the Yucca Mountain, Nevada, site for development as a geologic repository. On July 9, 2002, the U.S. Congress approved this recommendation in Joint Resolution 87, which designated Yucca Mountain as the repository for spent nuclear waste. On July 23, 2002, the President signed Joint Resolution 87 into law (Public Law 107-200, 116 Stat. 735 [2002]).

A nuclear fuel pellet the size of the tip of your little finger is equivalent to the energy provided by 1780 pounds of coal or 149 gallons of oil.

The DOE is currently (early-2006) focused on the third major step of the process. The DOE is preparing an application to obtain a license from the NRC to construct a repository. The NRC is responsible for developing the regulations to implement the U.S. Environmental Protection Agency (EPA) safety standards and for licensing the repository. After the DOE submits the license application to the NRC, the NRC has three years to review the application and could request a fourth year from Congress, if needed, to make its determination on licensing.

If licensed, the final step is construction and operation of the facility. The construction process would be conducted by the DOE, which has the responsibility for developing a permanent disposal facility for spent fuel and other high-level waste and which would operate the facility.

Approximately 40,000 metric tons of used fuel has been produced to date by the U.S. nuclear energy industry during more than 40 years of operation. If all the fuel assemblies were stacked side by side and laid end to end they would cover an area the size of a football field to a depth of about five yards.

4.5.3 If the repository is not yet completed, where is the spent nuclear fuel currently being stored?

Every 1 to 2 years, approximately one-third of the nuclear fuel in an operating reactor needs to be unloaded and replaced with new fuel. The used fuel is commonly called "spent nuclear fuel." Nuclear power facilities have temporary storage for spent fuel in steel-lined concrete pools that are filled with water (spent fuel pools). The water acts as a natural barrier for radiation from the fuel assemblies and keeps the fuel thermally cool while it decays and becomes less radioactive. Because the designers of the nuclear power facilities originally anticipated that the spent fuel would be reprocessed (see response to Question 4.5.9), they designed the nuclear facilities to store about a decade's worth of used fuel. However, at this time commercial reprocessing is not being pursued.

If the storage capacity of the spent fuel pool is approached, then licensees may consider alternatives, such as above-ground dry storage casks. In dry storage casks, spent fuel is surrounded by inert gas inside a sealed metal cylinder that is enclosed within a metal or concrete outer shell. Depending on the design of the casks, they are either placed horizontally or vertically on a concrete pad. The pad, casks, and associated security infrastructure are called an independent spent fuel storage installation (ISFSI). The NRC approves the design of the casks after conducting a technical review to ensure that the casks are safe and secure for use at nuclear power facilities. The NRC has approved 14 cask designs for use. By the beginning of 2006, there were ISFSIs at:

- 28 nuclear power reactor sites

- 8 decommissioned or decommissioning nuclear power reactor sites

- two storage facilities operated by the DOE at the Idaho National Environmental and Engineering Laboratory near Idaho Falls, Idaho

- one ISFSI at the General Electric Morris facility in Illinois.

Another option for storage of fuel is in an away-from-reactor interim storage facility. Private Fuel Storage, LLC (PFS), has submitted a request to the NRC for a license to build such a privately owned facility on the reservation of the Skull Valley Band of Goshute Indians, about 80 kilometers (50 miles) southwest of Salt Lake City, Utah. On February 24, 2005, the Atomic Safety and Licensing Board (ASLB) completed its review of the proposed spent nuclear fuel storage facility and ruled in favor of the PFS. The Commission upheld the ASLB decision in a Memorandum and Order dated September 9, 2005, and authorized the NRC staff to issue a license to construct and operate the PFS facility.

4.5.4 What will happen if Yucca Mountain is never finished or approved for storing nuclear waste?

The NRC's Waste Confidence Rule, found in 10 CFR 51.23, states that "the Commission has made a generic determination that, if necessary, spent fuel generated in any reactor can be stored safely and without significant environmental impacts for at least 30 years beyond the licensed life for operation (which may include the term of a revised or renewed license) of that reactor at its spent fuel storage basin or at either onsite or offsite independent spent fuel storage installations. Further, the Commission believes there is reasonable assurance that at least one mined geologic repository will be available within the first quarter of the twenty-first century, and sufficient repository capacity will be available within 30 years beyond the licensed life for operation of any reactor to dispose of the commercial high-level waste and spent fuel originating in such reactor and generated up to that time."

The staff is confident that there will eventually be a licensed high-level waste repository. If the site near Yucca Mountain is eventually found to be unsuitable, alternative sites will be considered. Until a permanent high-level waste repository is operational, the spent nuclear fuel will be safely stored either onsite or at offsite interim storage facilities.

4.5.5 Who is paying for the storage of spent fuel now and who will pay for the transportation to and storage of spent fuel at Yucca Mountain?

The storage of spent fuel onsite (either in a spent fuel pool or an independent spent fuel storage installation [ISFSI]) is paid for by the licensee and ultimately by electricity consumers. The transportation and disposal of spent fuel at a centralized repository (such as Yucca Mountain) is also funded by electricity consumers. The Nuclear Waste Policy Act established the Nuclear Waste Fund as a means to pay for a permanent repository, an interim storage facility (if needed), and the transportation of used fuel. Since 1982, electricity consumers have paid into the fund a fee of one-tenth of one cent for every nuclear-generated kilowatt-hour of electricity consumed. By the end of the twentieth century, customer payments plus interest totaled more than $16 billion.

4.5.6 How is an onsite storage facility licensed?

High-level Waste Shipment

Onsite storage facilities are licensed separately from the reactor license renewal process. The NRC authorizes storage of spent nuclear fuel at an independent spent fuel storage installation (ISFSI) under two licensing options: a site-specific license or a general license.

Under a site-specific license, an applicant submits a license application to NRC and the NRC performs a technical review of all the safety aspects of the proposed ISFSI. If the application is approved, the NRC issues a site-specific license that is valid for 20 years. The spent fuel storage license contains technical requirements and operating conditions (fuel specifications, cask leak testing, surveillance, and other requirements) and specifies what the licensee is authorized to store at the site. The site-specific license is a stand alone license, independent of the NRC license issued to possess and operate a nuclear power facility.

Independent Spent Fuel Storage Installation

A general license authorizes a nuclear power plant licensee to store spent fuel in NRC-approved casks at an existing site that is licensed for operating of a power reactor under 10 CFR Part 50. An NRC-approved cask is one that has undergone a technical review of its safety aspects and been found to meet all of the NRC's requirements in 10 CFR Part 72. The NRC issues a Certificate of Compliance for a cask design to a cask vendor after a rulemaking determines its technical adequacy. The cask certificate expires 20 years from the date of issuance. Licensees are required to perform evaluations of their sites to demonstrate that the site is adequate for storing spent fuel in dry casks. These evaluations must show that the cask Certificate of Compliance conditions and technical specifications can be met. The licensee must also review its security program, emergency plan, quality assurance program, training program, and radiation protection program, and make any necessary changes to incorporate the ISFSI at its reactor site.

4.5.7 Is the security of the nuclear waste stored onsite being reviewed?

Although it is very unlikely that any substantial radiological release would occur from a terrorist attack on a spent fuel pool or dry cask storage facility, the NRC is conducting a comprehensive evaluation, which includes consideration of potential consequences of terrorist attacks. Assessing the precise amount of contamination resulting from a release depends on many factors, such as type and amount of damage to the pool or dry cask storage facility, location of the damage, proximity of the storage facility to populated areas, and meteorological conditions at the time of the event. As part of this evaluation, the agency will consider the need for additional requirements to enhance licensee security and public safety.

4.5.8 Are onsite storage facilities secure from terrorist attacks?

The NRC considers spent fuel storage facilities to be robust; in the event of a terrorist attack similar to those of September 11, 2001, it is unlikely that any substantial radiological release would occur. Unlike the structures that were destroyed on September 11, 2001, spent fuel pools and dry storage casks are not constructed of flammable material that would fuel long-duration fires. If an attack were to occur, licensees have approved emergency plans, tested biennially, that coordinate local, state, and Federal government responses. The NRC believes that the health and safety of the public are well protected.

4.5.9 What is the policy of the United States concerning reprocessing?

Reprocessing of spent nuclear fuel involves the chemical treatment of the fuel to separate unused uranium and plutonium from radioactive fission products. Spent nuclear fuel can be reprocessed and some recovered plutonium can be used in new fuel assemblies. When most U. S. nuclear plants were built, the industry, with the Federal government's encouragement, planned to recycle or reprocess used nuclear fuel. In 1979, a decision was made by President Carter to ban commercial nuclear fuel reprocessing because of concerns about possible proliferation of weapons-grade material. President Regan lifted the reprocessing ban in 1981, however there was little or no interest in pursuing reprocessing by the nuclear industry. In early 2006 the U.S. DOE announced a new initiative called the Global Nuclear Energy Partnership which envisions the development and deployment of a closed fuel cycle that enables the recycling and consumption of long lived radioactive waste. Such a program would require the reprocessing of spent fuel. Reprocessing of spent fuel does occur in other countries.

4.6 Human Health Issues

The most commonly asked questions relating to human health issues include the potential for radiation exposure to the public and the potential for adverse effects from such exposure. This section responds to commonly asked questions regarding radiation exposure and its effect on human health.

4.6.1 What is radiation and where does it come from?

Radiation is naturally present in our environment and has been since the planet was formed. Radiation is a form of invisible energy waves or particles. It is emitted from unstable atoms as they change to become more stable. Such atoms are termed "radioactive" and materials containing significant amounts of radioactive atoms are called "radioactive material." Life has evolved in an environment that has significant levels of ionizing radiation. It comes from outer space (cosmic), the ground (terrestrial), and even from within our own bodies. It is present in the air we breathe, the food we eat, the water we drink, and the construction materials we use to build our homes. Certain foods such as bananas and brazil nuts naturally contain higher levels of radioactive material than other foods. Brick and stone homes have higher natural radiation levels than homes made of other building materials such as wood.

Radiation is a form of invisible energy waves or particles. It is emitted from unstable atoms as they change to become more stable.

During the late nineteenth century, scientists discovered natural radioactive elements. In the early twentieth century, scientists were able to create radioactive elements from stable elements. In 1942, scientists were able to split atoms deliberately, which released the energy that was in the nucleus and created unstable atoms in the process. Although there are different types of energy and particles emitted from different types of radioactive material, there is no difference between natural and man-made radiation.

Radiation dose is measured in a unit called a rem, which is based on the effect of radiation on the human body. It takes into account both the amount of radiation deposited in body tissues and the type of radiation. Radiation dose is often measured in millirem, or one-thousandth of a rem. In the International System of units (SI units) the unit of dose is the sievert (Sv), which is equivalent to 100 rems. The average person in the United States receives about 360 millirems of radiation a year (3.6 mSv per year). About 300 millirems (3 mSv) are from natural sources and 60 millirems (0.6 mSv) are from human-made sources.

Table 4.1 illustrates the dose that an average person receives annually (the annual effective dose equivalent), the sources of radiation, and the fractions of radiation exposure from various sources. Approximately 82 percent of our total exposure to radiation comes from natural sources, including radon gas (approximately 55 percent of our exposure to natural sources), the sun and outer space (8 percent), the earth's soil and rocks (8 percent), and the human body itself (11 percent). The remaining 18 percent of our total radiation exposure comes from human-made or artificial sources, primarily medical and dental x-rays and consumer products. The nuclear fuel cycle is responsible for less than 3/100ths of 1 percent of the total annual radiation dose to the average person (based on the calculated dose from all facets of the nuclear power cycle divided by the population of the United States).

Table 4.1. Annual Radiation Dose to an Average Individual (Annual Effective Dose Equivalent)

Source	Dose (mrem/yr)[a]	Percent of Total
Natural		
Radon	200	55
Cosmic rays	27	8
Terrestrial (soil and rocks)	28	8
Internal (body)	39	11
Total Natural	300	82
Human-made		
Medical x-ray	39	11
Nuclear medicine	14	4
Consumer products[b]	10	3
Occupational[c]	0.9	<0.3
Nuclear fuel cycle	<1	<0.03
Fallout	<1	<0.03
Miscellaneous	<1	<0.03
Total human-made	63	18
Total Natural and Human-made	363	100

Source: Adapted from NCRP Report 93, "Public Radiation Exposure from Nuclear Power Generation in the United States," as abstracted by the University of Michigan (http://www.umich.edu/~radinfo/)

(a) 1 mrem is equal to 0.01 millisieverts.

(b) Such as radon in domestic water supplies, building materials, mining, and agricultural products, and tobacco.

(c) Individual employed in occupations that utilize radioactive materials or sources of radioactivity such as nuclear medicine, manufacturing, and power production.

4.6.2   Is radiation harmful?

Health effects from exposure to radiation range from no effect at all to death and can be responsible for inducing diseases such as leukemia, breast cancer, and lung cancer. Very high (hundreds of times higher than a rem), short-term doses of radiation have been known to cause prompt (or early, also called acute) effects, such as vomiting and diarrhea, skin burns, cataracts, and even death.

Radiation is only one of many agents with the potential for causing cancer, and cancer caused by radiation cannot be distinguished from cancer attributed to other causes, such as chemical carcinogens.

When radiation interacts within the cells of our bodies, several events can occur. First, the damaged cells can repair themselves and permanent damage does not result; this is the most common outcome for x-rays, gamma radiation, and beta radiation. Second, the cells may die, much like large numbers of cells do every day in our bodies, and dead cells may be replaced through normal biological processes. Third, the cells may either incorrectly repair themselves, resulting in a change in the cells' genetic structure that can mutate and subsequently be repaired without any effect, or can sometimes form pre-cancerous cells that may become cancerous. Radiation is only one of many agents with the potential for causing cancer, and cancer caused by radiation cannot be distinguished from cancer attributed to other causes, such as chemical carcinogens.

The associations between radiation exposure and the development of cancer are mostly based on studies of populations exposed to relatively high levels of ionizing radiation (for instance, the Japanese atomic bomb survivors and the recipients of selected diagnostic or therapeutic medical procedures). Although radiation can cause cancers at high doses and high dose rates, currently there are no data to unequivocally establish the occurrence of cancer following exposures to low doses and dose rates below about 10 rems (10,000 millirems [0.1 sieverts]). For example, people living in areas of the country that receive greater levels of background radiation (such as Denver, Colorado) do not show higher rates of cancer.

People living in areas of the country that receive greater levels of background radiation (such as Denver, Colorado) do not show higher rates of cancer.

The chances of getting cancer from a low dose of radiation is not known precisely because the few effects that may occur cannot be distinguished from normally occurring cancers. The normal chance of dying from cancer is about one in five. The actual amount of radiation any member of the public receives from activities occurring at nuclear power facilities is so small that scientists have been unable to make empirically based estimates of radiation risk from such low levels of exposure with any precision.

There are many difficulties involved in designing research studies that can accurately measure the projected small increases in cancer cases that might be caused by low exposures to radiation when compared to the normal rate of cancer. The best that scientists can do is to make an unsubstantiated assumption that any amount of radiation may pose some risk for causing cancer or having some hereditary effect and that the risk is higher for higher radiation exposures. This is called a linear, no-threshold dose response model and is used to describe the relationship between radiation dose and the occurrence of cancer. It is known that this model errs on the side of overestimating radiation risks. This model suggests that any increase in dose above background levels, no matter how small, results in an incremental increase in risk above existing levels of risk. Although the NRC has accepted this hypothesis as a conservative (i.e., cautious) model for determining radiation standards, the NRC, like other authoritative bodies, recognizes that this model probably overestimates radiation risk.

The NRC sets limits on radiological effluents, requires monitoring of effluents and foodstuffs to ensure that those limits are met, and has set dose limits to regulate the release of radioactive material from nuclear power facilities.

4.6.3 How much radiation is released from a nuclear power facility?

The NRC has established strict limits on the amount of radioactive releases to the environment allowed from nuclear power facilities and the resulting exposure for members of the public. These requirements are given in 10 CFR Part 20, Appendix B, Table 2 (http://www.nrc.gov/reading-rm/doc-collections/cfr/part020/part020-appb.html). Whereas contaminants may be present and detectable offsite, the release limits have been designed and proven to be protective of the health and safety of the public (including sensitive populations) and the environment.

The NRC sets limits on radiological effluents, requires monitoring of effluents and foodstuffs to ensure that those limits are met, and has set dose limits to regulate the release of radioactive material from nuclear power facilities. All reactor licensees monitor their effluents and calculate offsite doses caused by radioactive liquid and gaseous effluents and direct radiation. These calculations are performed to demonstrate the licensee's compliance with its technical specifications and NRC regulations. Requirements for redundancy in monitoring as well as the monitoring of various pathways that could result in the release of radiation to the environment ensure that unmonitored and unplanned releases are avoided. The licensee's Offsite Dose Calculation Manual (ODCM) provides for collection and analysis of a variety of samples such as soil, water, plants, and animals. Actual measurements are made of the liquid and airborne releases from the facility, and they are verified by the monitoring program described in the ODCM. As a result of these criteria, the average person (not

An NCI study concluded that there is "no evidence that an excess occurrence of cancer has resulted from living near nuclear facilities."

including a radiation worker employed at the facility) living within 80 kilometers (50 miles) of a nuclear power facility receives less than 1 millirem per year (0.01 millisieverts per year) of radiation dose from the nuclear power facility. This is compared to the approximately 300 millirems per year (3 millisieverts per year) received from natural sources and 60 millirems per year (0.6 millisieverts per year) from human-made sources, as discussed in the response to Question 4.6.1. This dose can also be compared to the radiation received from the earth's crust, which ranges from 23 millirems per year (0.23 millisieverts per year) along the Atlantic Coast to 90 millirems per year (0.9 millisieverts per year) on the Colorado Plateau. Other sources of radiation that are common in our lives include airline flights, which give about 1 millirem (0.01 millisieverts) of radiation dose per 1,600 kilometers (1,000 miles) flown. A round-trip cross-country airplane trip would give a dose of about 5 millirems (0.05 millisieverts). The dose from watching television is about 1-2 millirem (0.01-0.02 millisieverts) per year, and from a single medical x-ray is about 40 millirems (0.4 millisieverts).

4.6.4   Does radiation from nuclear power facilities cause cancer?

The average annual dose to a member of the public from a nuclear power facility is in the range of less than 1/1000th rem (1 millirem) per year (0.01 millisieverts per year). This is compared to the 10 rems (10,000 millirems [100 millisieverts]) discussed in the response to Question 4.6.2. At doses above 10 rem (0.1 Sv) a relationship between radiation and health effects can be observed. There are no data to unequivocally establish the occurrence of health effects or cancer following exposures to low doses and dose rates below 10 rem (0.1 Sv). Although there is a statistical chance that radiation levels that small could result in a cancer, it has not been possible to calculate with any certainty the probability of receiving cancer from a dose this small. Because many agents cause cancer, it is often not possible to say conclusively whether it is a radiation-induced cancer or not. At the request of Congress, the National Cancer Institute (NCI) published a study in 1991, "Cancer in Populations Living Near Nuclear Facilities," which looked at cancer mortality rates around 52 nuclear power facilities, 9 U.S. Department of Energy facilities, and 1 former commercial fuel reprocessing facility. The NCI study concluded that there is "no evidence that an excess occurrence of cancer has resulted from living near nuclear facilities." Additionally, the American Cancer Society has concluded that although reports about cancer case clusters in such communities have raised public concern, studies show that clusters do not occur more often near nuclear plants than they do elsewhere in the population.

The American Cancer Society has concluded that although reports about cancer case clusters in such communities have raised public concern, studies show that clusters do not occur more often near nuclear plants than they do elsewhere in the population.

4.6.5   I have read reports stating that there are excess cases of a specific type of cancer in the vicinity of a specific nuclear facility. Doesn't that mean that radiation from nuclear power facilities causes cancer?

Authors of various reports have stated or implied that there are cause-and-effect relationships in the statistical associations between cancer rates and reactor operations. While it is true that cancer rates vary among locations, it is very difficult to ascribe the cause of a cluster of cancers to some local environmental exposure, such as radiation from a nuclear power facility. Statistical association alone does not demonstrate causation, and well-established scientific methods must be used to determine that for two things that appear to be associated over time, it can be concluded that one causes the other. For example, a person could say, "In the winter I wear boots, and in the winter I get colds." While there is a strong statistical association between wearing boots and getting colds, it would be inappropriate to say that wearing boots causes colds.

The scientific community adheres to several principles of good science that need to be employed before a cause-and-effect claim can be made. These principles include whether the study can be replicated, whether it has considered all the data or was selective (e.g., in the population or in the years studied), whether it evaluated all possible explanations for the observations, whether the data were valid and reliable, and whether its conclusions were subjected to independent peer review, evaluation, and confirmation.

A number of studies that conformed to these principles have been performed to examine the health effects around nuclear power facilities:

- National Cancer Institute – In 1990, at the request of Congress, the National Cancer Institute conducted a study of cancer mortality rates around 52 nuclear power plants and 10 other nuclear facilities. The study covered the period from 1950 to 1984 and evaluated the change in mortality rates before and during facility operations. The study concluded there was no evidence that nuclear facilities may be linked causally with excess deaths from leukemia or from other cancers in populations living nearby.

- University of Pittsburgh – Investigators from the University of Pittsburgh found no link between radiation released during the 1979 accident at the Three Mile Island nuclear station and cancer deaths among nearby residents. Their study followed for a period of 20 years over 32,000 people who lived within 8 kilometers (5 miles) of the facility at the time of the accident.

- Connecticut Academy of Sciences and Engineering – In January 2001, the Connecticut Academy of Sciences and Engineering issued a report on a study around the Haddam Neck nuclear power plant in Connecticut and concluded that exposures to radionuclides were so low as to be negligible and found no meaningful associations to the cancers studied.

- American Cancer Society – In 2004, the American Cancer Society concluded that although reports about cancer clusters in some communities have raised public concern, studies show that clusters do not occur more often near nuclear plants than they do by chance elsewhere in the population. Likewise, there is no evidence that links the isotope strontium-90 with increases in breast cancer, prostate cancer, or childhood cancer rates. Radiation emissions from nuclear power plants are closely controlled and involve negligible levels of exposure for nearby communities.

- Florida Bureau of Environmental Epidemiology – In 2001, the Florida Bureau of Environmental Epidemiology reviewed claims that there are striking increases in cancer rates in southeastern Florida counties caused by increased radiation exposures from nuclear power plants. However, using the same data to reconstruct the calculations on which the claims were based, Florida officials were not able to identify unusually high rates of cancers in these counties compared with the rest of the state of Florida and the nation.

- Illinois Public Health Department – In 2000, the Illinois Public Health Department compared childhood cancer statistics for counties with nuclear power plants to similar counties without nuclear plants and found no statistically significant difference.

Field Radiation Measurements

In summary, there are no studies to date that are accepted by the scientific community that show a correlation between radiation dose from nuclear power facilities and cancer incidence in the general public. The amount of radioactive material released from nuclear power facilities is well measured, well monitored, and known to be very small. The doses of radiation that are received by members of the public as a result of exposure to nuclear power facilities are so low that resulting cancers have not been observed and would not be expected.

4.6.6 How are radiation and releases of radioactive material regulated and monitored at nuclear power facilities?

NRC regulations require licensees to control and limit releases to the environment (the air and water) to very small amounts. As part of NRC requirements for operating a nuclear power facility, licensees must keep releases of radioactive material to unrestricted areas during normal operation as low as reasonably achievable (as described in the NRC's regulations in 10 CFR Part 50.36a) and comply with radiation dose limits for the public as given in the regulations in 10 CFR Part 20.

Wildlife Refuge

In addition, NRC regulations require licensees to maintain various effluent and environmental monitoring programs so that the impacts from plant operations are minimized and the extent of releases are accurately recorded and reported.

The control of releases is accomplished by barriers. One method used to control the release of radioactive material to the environment is to keep contaminated areas of the plant under negative pressure so that air leaks into the building, rather than out. In addition, exhaust pathways out of the building may be filtered to prevent the movement of radionuclides into the environment. Exhaust pathways are monitored so that there is a proper characterization of material that may be leaving the plant. Workers in contaminated areas are also monitored, along with any tools or equipment that is moved from the building, in order to prevent the spread of radioactive material.

The NRC requires licensees to report plant discharges and results of environmental monitoring around their plants to ensure that potential impacts are detected and reviewed. Licensees must also participate in an interlaboratory comparison program, which provides an independent check of the accuracy and precision of environmental measurements.

Licensees are required to keep accurate records on releases to the air and water. In annual reports, licensees identify the amount of liquid and airborne radioactive effluents discharged from plants and calculate associated doses. Licensees also must report environmental radioactivity levels around their plants annually. These reports, which are available to the public, include sampling from thermoluminescent dosimeters (which measure radiation dose levels); airborne radioiodine and particulate samplers; samples of surface, groundwater, and drinking water and downstream shoreline sediment from existing or potential recreational facilities; and samples of ingestion sources such as milk, fish, invertebrates, and broad-leaf vegetation.

The NRC conducts periodic onsite inspections of each licensee's effluent and environmental monitoring programs to ensure compliance with NRC requirements.

The NRC conducts periodic onsite inspections of each licensee's effluent and environmental monitoring programs to ensure compliance with NRC requirements. The NRC documents licensee effluent releases and the results of their environmental monitoring and assessment effort in inspection reports that are available to the public.

Over the past 25 years, radioactive effluents released from nuclear power facilities have decreased significantly. During the early part of that period, a significant contributor to the reduction was the addition of special systems (called augmented offgas systems) to boiling water reactors, which process some of the noncondensible gases formed in the reactor process to limit the radioactive gases released to the environment. In recent years, improved fuel performance and licensees' improved effluent control programs further contributed to reducing radioactive effluents.

Dose levels to the public during the license renewal period are not expected to increase from those during the initial licensing period.

4.6.7 Will radiation dose rates to the public increase during the license renewal period?

NRC regulations contain criteria and requirements for nuclear power plant licensing that ensure an acceptable level of plant safety, i.e., an acceptably low level of risk to public health and safety. The regulations are based on sound engineering precepts that are judged to be acceptable for safe plant design and operation. Dose levels to the public during the license renewal period are not expected to increase from those during the initial licensing period. No aging phenomenon has been identified that is expected to increase public radiation doses. Data obtained from measurements near nuclear power facilities suggests that, if anything, radiation doses to the public related to commercial nuclear power operation are decreasing.

At the low radiation doses from nuclear power plant operation, it is highly unlikely that any deaths will occur as a result of 20 years of additional operation of a nuclear power reactor.

4.6.8 What are the radiological health implications of extending the license for a reactor for 20 years?

According to 10 CFR Part 51, Subpart A, Appendix B, Table B-1, "the 100 year environmental dose commitment to the U.S. population from the fuel cycle, high level waste and spent fuel disposal excepted, is calculated to be about 14,800 person-rem (148 person-sieverts) or 12 cancer fatalities, for each additional 20 year power reactor operating term."

This calculated value of 12 additional deaths from fatal cancer over the 20 years of additional operation of a nuclear power plant is based on several very conservative assumptions. This calculated value does not represent real expected deaths. Realistically, no deaths are expected. At the low radiation doses from nuclear power plant operation, it is highly unlikely that any deaths will occur as a result of 20 years of additional operation of a nuclear power reactor.

These calculations use the concept of collective dose, which estimates the effects of radiation dose across a very large population. It assumes that a small amount of radiation dose spread out among a large population would yield effects similar to a larger amount of radiation dose to a much smaller population. This is intentionally a very conservative assumption, i.e., it estimates a dose greater than what could be reasonably expected of actual situations. According to the Health Physics Society (www.hps.org), "[b]elow the dose of ten rem, estimations of adverse health effect is speculative. Collective dose remains a useful index for quantifying dose in large populations and in comparing the magnitude of exposure from different radiation sources. However, for a population in which all individuals receive lifetime doses of less than 10 rem above background, collective dose is a highly speculative and uncertain measure of risk and should not be quantified for the purposes of estimating population health risks."

The cancer risk factors used in this calculation are also quite conservative. They are from the 1990 BEIR-V report of the National Research Council's Committee on the Biological Effects of Ionizing Radiation, *Health Effects of Exposure to Low Levels of Ionizing Radiation*. In this report, it is estimated that "if 100,000 persons of all ages received a whole body dose of 0.1 gray (10 rad) of gamma radiation in a single brief exposure, about 800 extra cancer deaths would be expected to occur during their remaining lifetimes in addition to the nearly 20,000 cancer deaths that would occur in the absence of radiation. Because the extra cancer deaths would be indistinguishable from those that occurred naturally, even to obtain a measure of how many extra deaths occurred is a difficult statistical estimation problem."

A recent report by the National Research Council (2009), BEIR VII Phase II, "Health Risks from Exposure to Low Levels of Ionizing Radiation" supports the linear, no-threshold dose risk model which suggests that the risk of cancer proceeds in a linear fashion at lower doses without a threshold and that the smallest dose has the potential to cause a small increase in risk to humans.

U.S. Environmental Protection Agency (EPA) regulation 40 CFR 190 limits doses to individuals in the public to 25 millirems (0.25 millisieverts) per year to the whole body. No one living near a nuclear power plant in the United States receives a dose above this limit, and very few receive a dose above 1 millirem per year (0.01 millisieverts per year), one-thousand times less than 1 rem (0.01 sieverts) and ten-thousand times less than the 10-rad (0.1 sieverts) value (1 rem is approximately equal to 1 rad [0.01 gray] for radiation from nuclear power reactors) used in the BEIR-V calculation.

The total radiation dose contribution to the population from current nuclear power plants is estimated to be 4.8 person-rem (0.048 person-sieverts) per year, while the contribution to the population from the complete uranium fuel cycle is 136 person-rem (1.36 person-sieverts) per year. The dose to an individual is only a very small fraction of these population doses.

The total annual release of strontium 90 into the atmosphere from all U.S. nuclear power plants is typically 1/1,000th of 1 curie, which is so low that the only chance of detecting strontium 90 is sampling the nuclear power plant effluents themselves.

4.6.9 Have there been studies showing an increase in strontium-90 radiation levels in baby teeth and corresponding cancer incidence as a result of releases of radioactive material from nuclear power plants?

In 2000, a report entitled *Strontium-90 in Deciduous Teeth as a Factor in Early Childhood Cancer* was published by the Radiation and Public Health Project. The report alleges that there has been an increase in cancer incidence due to strontium-90 released from nuclear power facilities. Elevated levels of strontium-90 in deciduous (baby) teeth was claimed in the report as the evidence for the increase in childhood cancer.

The National Environmental Policy Act requires the consideration of alternatives to the proposed action in an environmental impact statement (EIS). Reasonable alternatives include those that are practical or feasible from the technical and economic standpoint.

There are three sources of strontium-90 in the environment: fallout from nuclear weapons testing, releases from the Chernobyl accident in the Ukraine, and releases from nuclear power reactors. The largest source of strontium-90 is from weapons testing fallout as a result of above-ground explosions of nuclear weapons (approximately 16.9 million curies of strontium-90). The Chernobyl accident released 216,000 curies of strontium-90.

The total annual release of strontium-90 into the atmosphere from all U.S. nuclear power plants is typically 1/1,000th of 1 curie, which is so low that the only chance of detecting strontium-90 is sampling the nuclear power plant effluents themselves. The NRC regulatory limits from effluent releases and subsequent doses to the public are based on the radiation protection recommendations of international and national organizations such as the International Commission on Radiological Protection (ICRP) and the NCRP. Gaseous effluent releases are monitored at nuclear power facilities, and the results of monitoring performed by the licensees are reported annually to the NRC. The effluent release program and the licensee's monitoring programs are reviewed during the environmental review of license renewal.

In a report published in 2001, the American Cancer Society concluded that although reports about cancer case clusters in communities surrounding nuclear power plants have raised public concern, studies show that clusters do not occur more often near nuclear plants than they do by chance elsewhere in the population. The National Council on Radiation Protection and Measurements (NCRP) has found no statistically significant excess of biological effects due to strontium-90 exposures at levels characteristic of worldwide fallout, which is the greatest source of strontium-90 in the environment. Likewise, there is no new evidence that links strontium-90 with increases in breast cancer, prostate cancer, or childhood cancer rates. The American Cancer Society recognizes that public concern about environmental cancer risks often focuses on risks for which no carcinogenicity has been proven or on situations where known exposures to carcinogens are at such low levels that risks are negligible. The report states that "ionizing radiation emissions from nuclear facilities are closely controlled and involve negligible levels of exposure for communities near such plants."

Wind Generation

4.7 Alternatives

The site-specific supplements to the generic environmental impact statement on license renewal (SEISs) contain a chapter related to alternatives to the proposed action. A consideration of these alternatives is required by National Environmental Policy Act (NEPA). This section contains responses to questions regarding the selection and consideration of alternatives.

4.7.1 Why does the NRC consider alternatives to license renewal?

The National Environmental Policy Act (NEPA) requires the consideration of alternatives to the proposed action in an environmental impact statement (EIS). The President's Council on Environmental Quality (CEQ) says that "reasonable alternatives include those that are practical or feasible from the technical and economic standpoint." It also states that the alternatives are developed "using common sense rather than [being] simply desirable from the standpoint of the applicant." NEPA also requires the alternatives analysis in the EIS to "include the alternative of no action." The interpretation of "no action" depends on the nature of the proposal being evaluated. In the case of license renewal, the "no action" alternative may be thought of in terms of continuing with the present course of action, i.e., operation under the current operating license until the license has expired. Once the license has expired, the licensee must begin decommissioning the facility.

> *The NRC's responsibility is to ensure the safe operation of nuclear power facilities and not to formulate energy policy or encourage or discourage the development of specific alternative power generation.*

4.7.2 Doesn't conducting a review and granting the license so far in advance of the termination of the current license almost guarantee that an alternative will not be seriously considered by the licensee?

If a licensee is going to choose an alternative to license renewal, it would want the replacement facility (if one is to be built) to be ready to produce power by the end of the operating term of the nuclear power facility. In many cases, it can take up to 10 years to design and construct major new generating facilities. Thus, conducting the review 10 to 20 years in advance of the termination of the current license is not unreasonable for an applicant that may be required to use an alternative to license renewal.

4.7.3 Why doesn't the NRC encourage green alternative energy sources such as solar or wind power?

The NRC's responsibility is to ensure the safe operation of nuclear power facilities and not to formulate energy policy or encourage or discourage the development of specific alternative power generation. The staff's evaluation of alternatives in an environmental impact statement is limited to assessing their environmental impact rather than recommending energy alternatives.

Solar Panels

4.7.4 If an alternative is found that clearly has less environmental impact, why doesn't the NRC require the licensee to pursue the alternative?

The NRC's requirements to consider the environmental impacts of various alternatives is based on the National Environmental Policy Act (NEPA) of 1969. The purpose of NEPA is to ensure that relevant agencies examine and disclose the potential environmental impacts of their actions before taking the action. NEPA is a procedural statute that does not dictate a decision based on relative environmental impacts. Furthermore, the NRC has no authority or regulatory control over the ultimate selection of future energy alternatives. Likewise, the NRC cannot ensure that environmentally superior energy alternatives are used in the future. The NRC makes a decision to renew or not to renew a license based on safety and environmental considerations. The final decision on whether or not to continue operating the nuclear facility will be made by the licensee and by state and Federal (non-NRC) decision-makers (see response to Question 1.2.10). This final decision will be based on economics, energy reliability goals, and other objectives over which the other entities may have jurisdiction. Moreover, given the absence of the NRC's authority in the general area of energy planning, the NRC's identification of a superior alternative does not guarantee that such an alternative will be used.

As a result, based on the uncertainties involved and the lack of control that the NRC has in the choice of energy alternatives in the future, the Commission decided to exercise its NEPA authority to reject license renewal applications only in cases where there is such an imbalance between the impacts of license renewal and the impacts of the alternatives that it would be unreasonable to allow further consideration of license renewal.

4.8 Security

Although security issues are considered to be outside the scope of the environmental evaluation for license renewal, one of the most common questions related to security is answered in this section.

4.8.1 What has the NRC done to improve security as a result of the terrorist attacks on 9/11?

Access Control Terminal

Prior to September 11, 2001, the security measures in place provided reasonable assurance that the health and safety of the public would be protected in the event of an attack that involved radiological sabotage. The security measures were designed to protect against the threats described in 10 CFR 73.1. However, since September 11, 2001, the defensive capability of the nuclear power industry has been significantly enhanced. The NRC issued orders requiring security enhancements, conducted a three-phase audit of licensees' security programs in the weeks following the terrorist attacks, improved the process for conducting background investigations of new employees at nuclear power facilities, and initiated a number of studies related to the protection of nuclear material and facilities. The NRC also initiated a number of studies on the effects of a crash of a large commercial aircraft into a nuclear power plant. The NRC has also issued more than 60 advisories to its licensees describing changes in the threat environment and providing guidance on ways to enhance security.

NRC major actions since September 11, 2001, have included the following:

- ordering plant owners to increase physical security to defend against a more challenging adversarial threat,

- requiring strict site access controls for personnel,

- requiring licensees to conduct vehicle checks at greater stand-off distances,

- improving liaison with Federal, state, and local agencies responsible for protection of the national critical infrastructure through integrated response planning,

Security Exercise

- enhancing communication and liaison with the intelligence community,

- improving communication between military surveillance authorities, the NRC, and its licensees to prepare power plants and to effect safe shutdown should it be necessary,

- ordering plant operators to improve their capability to respond to events involving explosions or fires,

- enhancing readiness of security organizations by strengthening training and qualification programs for plant security forces,

- enhancing force-on-force exercise to provide a more realistic test of plant capabilities to defend against an adversary force, and

- working with national experts to predict the realistic consequences of terrorist attacks on nuclear facilities, including one from a large commercial aircraft. For the facilities analyzed, the results confirm a low likelihood both for damaging the reactor core and releasing radioactivity that could affect public health and safety. Even in the unlikely event of a radiological release due to a terrorist use of a large aircraft against a nuclear power plant, the studies indicate that there would be time to implement the required onsite mitigating actions. These results have also validated the offsite emergency planning basis.

In addition, the NRC works with a variety of other Federal agencies, in particular the Department of Homeland Security and the Homeland Security Council, to ensure that security around nuclear power plants is well coordinated and that responders are prepared if a significant event occurs. If an event were to occur, the NRC would coordinate the resources of more than 18 Federal agencies in response to any radiological emergency.

5.0 Public Involvement in the License Renewal Process

Public involvement is a very important part of the environmental evaluation for license renewal. This section discusses the means by which members of the public may participate in the environmental evaluation. It also provides guidance for members of the public to access the documents on which the NRC's evaluation is based.

5.1 Public Involvement

5.1.1 How does a member of the public know that a licensee is planning to renew its license?

The public is notified through the *Federal Register*, press releases, and local advertisements after an application for the license renewal of a nuclear power plant has been received by the NRC. A notice is routinely placed in the *Federal Register* within a month after receipt of the application.

Licensees notify the NRC of their plans to submit an application for license renewal often years in advance of their submittal. The advance notice assists the NRC in workload planning and ensures that staff will be assigned to review the application upon its arrival. Members of the public can view the list of anticipated license renewal applications and the date that the licensees are anticipating sending the applications to the NRC. This list is provided on the NRC's website: http://www.nrc.gov/reactors/operating/licensing/renewal/applications.html.

5.1.2 Where do I find information related to license renewal for a specific nuclear power facility?

The status of license renewal activities and industry activities can be found on the NRC website at http://www.nrc.gov/reactors/operating/licensing/renewal/applications.html The following information is added to the web site when available:

The status of license renewal activities and industry activities can be found on the NRC website.

- contact information for the NRC license renewal safety and environmental project managers,

- a copy of the application,

- the license renewal review schedule,

- a list of meetings that are open to members of the public along with the agenda for the meetings,

- a transcript or meeting summary (as appropriate), copies of slides that were used at the meeting, or copies of inspection reports if pertinent,

- the draft and final site-specific supplement to the generic environmental impact statement on license renewal (SEIS), and

- the safety evaluation report.

5.1.3 What are the kinds of meetings that the public can be involved in and how does the public find out about them?

NRC Public Meeting

The NRC defines three categories of public meetings that are held for different purposes and with varying degrees of public participation. All three types of meetings are held during the license renewal process. The first type of meeting (Category 1) is commonly held with the applicant for a specific plant. Category 1 meetings provide the public with an opportunity to observe NRC's interactions with the applicant, to obtain information that assists the public in understanding regulatory issues, and to offer constructive comments. The public is invited to observe the meeting and has the opportunity to communicate with the NRC staff before the end of the meeting. Although most questions can be answered at the meeting, some questions require informal follow-up by telephone or email. Meetings held with the applicant to discuss the license renewal application are also considered Category 1 meetings.

Category 2 meetings are typically held with a group of representatives of industry, licensees, vendors, or non-government organizations, such as public interest and citizen groups, and focus on issues that could apply to several facilities. Meetings held during development of the generic environment impact statement for license renewal (GEIS) were Category 2 meetings.

Category 3 meetings are typically held with representatives of non-government organizations, private citizens or interested parties, or various businesses or industries. For Category 3 meetings, public participation is actively sought, with the intended objective being to provide a range of views, information, concerns, and suggestions about regulatory issues. This type of meeting provides the public the widest participation opportunities. The public scoping meetings on a site-specific supplement or the public meeting to discuss the draft site-specific supplement to the generic environmental impact statement on license renewal (SEIS) are considered Category 3 meetings.

For a typical license renewal application, a number of meetings are open to the public:

- meetings with the applicant to provide the NRC staff an overview of the license renewal application – Category 1,

- a meeting with members of the public to discuss the conduct of the safety review – Category 3,

- meetings with the applicant and members of the public to discuss the environmental review scoping process – Category 3,

- meetings with the applicant to present the results of the NRC's initial safety inspection (Scoping Methodology) of the company's license renewal program – Category 1,

- meetings with the applicant to present the results of the NRC's second safety inspection program (Aging Management Review) of the company's license renewal program – Category 1,

- meetings with the applicant to discuss potential open items that were identified by the NRC staff as part of its review – Category 1,

- meetings with the applicant and members of the public to discuss the draft and receive comments on the draft SEIS – Category 3,

- meetings with the applicant to discuss issues related to the safety evaluation report – Category 1, and

- a meeting with the Advisory Committee on Reactor Safeguards to discuss license renewal for a specific facility – Category 1.

5.1.4 What are the opportunities for public participation during the environmental review of the license renewal application?

Although public involvement and comments are invited and encouraged throughout the environmental review for a particular site, the public is specifically invited and encouraged to provide input at two critical stages during the environmental review of the license renewal application. The first stage is during the scoping process for the site-specific supplement to the generic environmental impact statement on license renewal (SEIS). This begins approximately 3 months after the applicant has submitted its application for license renewal. The public is notified at the beginning of the scoping process through the publication of a *Federal Register* notice, a meeting notice on the NRC website, through advertisements placed in local newspapers in communities near the nuclear power facility, and by flyers distributed throughout the local community. The scoping process is conducted to define the proposed action, to determine the scope of the SEIS, and to identify the significant issues to be analyzed in depth. The NRC website, *Federal Register* notice, and advertisements provide addresses for written comments to be submitted in person, by mail, or electronically. In addition, the notice contains the time and location of two public scoping meetings that are held in the vicinity of the nuclear plant. The meetings are commonly held on the same day, with one scheduled in the afternoon and the other during the evening to encourage a greater number of attendees. A description of the public scoping meetings is found in the response to Question 5.1.5. Scoping comments can also be given orally or submitted in writing at the public meetings. The deadline for scoping comments is usually 60 days following the publication of the notice in the *Federal Register*.

The public is specifically invited and encouraged to provide input at two critical stages during the environmental review of the license renewal application; during the scoping process for the site specific supplement, and following the publication of the draft SEIS.

After the comments have been received, they are evaluated and considered in the preparation of the site-specific analysis, as appropriate. The comments considered to be in scope are listed in Appendix A of the draft SEIS, along with the NRC staff's decision about whether the comment will be further evaluated as part of the analysis during the preparation of the draft SEIS.

The second opportunity for public participation occurs following the publication of the draft SEIS, which occurs approximately a year after the application is received. The NRC staff places a "notice of availability" in the *Federal Register* (and on the NRC website) with

Public Meeting for North Anna Power Station in Virginia

instructions for the public and other interested parties of how to obtain copies. A copy of this notice, along with a copy of the draft SEIS, is also sent to those persons attending the public scoping meeting who place their names on a list to receive further information on the license renewal process for that specific plant. The notice requests comments on the draft SEIS and provides addresses for delivering and sending the comments to the appropriate NRC staff member by mail or electronically. A 75-day period is allotted for the public's review and the receipt of comments.

Two public meetings are held near the nuclear plant to provide an overview of the draft SEIS and to accept additional public comments on the document. Again, the meetings are held on the same day, with one scheduled for the afternoon and the other for the evening. Every comment received is considered and, if appropriate, incorporated into the final document. All of the comments on the draft SEIS are listed in Appendix A of the final SEIS, along with the NRC staff's decision about whether the comment was within the scope of license renewal and, if appropriate, where changes to the text of the final SEIS were made in response to the comment.

5.1.5 What happens during the environmental public meetings that are held during the license renewal review process?

Site Audit

As discussed in the response to Question 5.1.4, there are two sets of public meetings for the environmental review of the license renewal application for each specific site. The purpose of the first set of meetings is to allow the public to participate in the scoping process. The purpose of the second set of meetings is to elicit public comments on the draft site-specific supplement. Each meeting begins with an open house, during which the public can view posters or displays related to the license renewal process. During this time, members of the public can informally discuss issues related to the license renewal review, including issues that are outside the scope of the environmental and safety reviews with the NRC staff, any applicant representative present, and any state or local officials attending the open house. At the scheduled time, the formal portion of the meeting starts with presentations by the NRC staff and its consultants to explain the license renewal process and to discuss the preparation and results of the draft site-specific supplement to the generic environmental impact statement on license renewal (SEIS). Frequently, the applicant also makes a presentation. Following the presentations, the NRC staff requests comments and questions from members of the public. The majority of the formal portion of the meeting is devoted to receiving comments and statements from members of the public. The NRC staff also accepts written or oral statements relating to the license renewal process.

When can I submit written or electronic comments and concerns during the review?

Although public involvement and comments are invited and encouraged throughout the environmental review for a particular site, the NRC solicits both written and oral comments from members of the public at two different times during the review. The first period of time is during the scoping process (see to identify significant issues to be analyzed in depth. Public scoping meetings are held near the nuclear plant that is seeking license renewal. Members of the public are invited to provide comments orally or in writing during these meetings. The NRC staff publishes a *Federal Register* notice that provides the times and locations. The notice is also placed in newspapers in communities near the plant and is posted on the NRC's website for the specific plant undergoing review. It provides addresses for written comments to be submitted in person, by mail, or electronically. The deadline for comments is usually 60 days following the publication in the *Federal Register* of the notice of intent to conduct scoping.

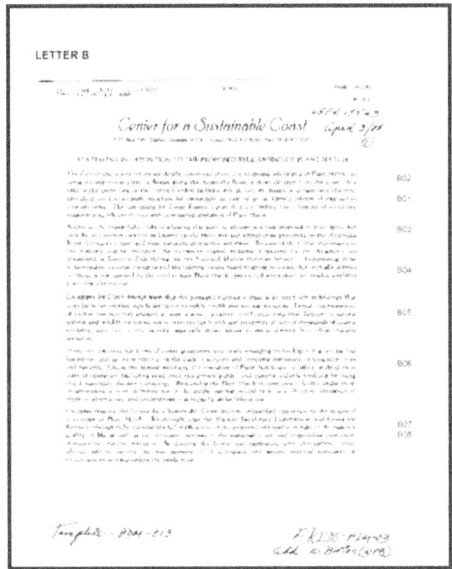

Short Comment Letter

The NRC also solicits written comments from members of the public following publication of the draft SEIS. The NRC staff places a notice in the *Federal Register* and on the NRC website that the draft SEIS has been issued with instructions for the public and other interested parties on how to obtain copies. Copies of the draft SEIS are also available on the NRC website or can be obtained as discussed in the response to Question 5.2.8. A copy of the notice and the draft SEIS is also sent to those people from the first meeting who requested a copy. The notice requests comments on the draft supplement and provides addresses for delivering or sending the comments to the appropriate NRC staff member. Usually, a 75-day period is allotted for the public's review and the receipt of comments. The NRC then holds a second set of public meetings in the vicinity of the nuclear facility to present the results of the draft SEIS to the public and to obtain comments, both oral and written, from the public.

5.1.7 Does NRC do anything to ensure that the public that opposes nuclear power knows about the review?

The NRC attempts to notify all stakeholders of any upcoming reviews. This includes Federal, state, and local agencies, as well as licensee staff, and members of the public or citizen advocacy groups that have previously expressed an interest in the regulatory activities related to a specific nuclear power facility. This also includes members of the public and organizations that oppose nuclear power. In addition to notices placed in the *Federal Register* or in local newspapers, the NRC staff notifies stakeholders (including members of the public or representatives of groups) who have previously attended public meetings related to a specific nuclear power facility or to license renewals. Frequently, these groups also receive a courtesy phone call to ensure they have been notified of public meetings on scoping and the preliminary conclusions in the draft site-specific supplement to the generic environmental impact statement on license renewal (SEIS).

5.1.8 Why doesn't the NRC hold a full adjudicatory hearing for each plant that requests license renewal?

Hearings on license renewal applications are not mandatory; that is, hearings are held only if a petition that shows standing to intervene and sets forth at least one contention (issue) that is suitable for litigation in the proceeding is filed.

5.1.9 As a member of the public how do I request intervention in the license renewal process? What is the timetable?

Any person whose interest may be affected by the proceeding to grant a renewed license to a specific facility may file a written request for a hearing or a petition for leave to intervene with respect to the renewal of the license. The regulations related to intervening in a licensing action are governed by 10 CFR 2.309.

Any person whose interest may be affected by the proceeding to grant a renewed license to a specific facility may file a written request for a hearing or a petition for leave to intervene with respect to the renewal of the license.

When the NRC receives a license renewal application, it is made available on the NRC's website at www.nrc.gov. Approximately 2 months after the NRC receives the application, a notice is posted in the *Federal Register* indicating the opportunity for a hearing regarding the renewal of the operating license and instructions for filing a request for a hearing or a petition for leave to intervene. Members of the public have a minimum of 60 days from the date of the *Federal Register* notice in which to file a request for a hearing or a petition for leave to intervene. A request for a hearing or petition for leave to intervene must be filed with the Secretary of the Commission, U.S. Nuclear Regulatory Commission, Washington, DC 20555-0001, Attention: Rulemaking and Adjudications Staff, or it may be delivered to the Commission's Public Document Room, 11555 Rockville Pike (first floor), Rockville, Maryland 20855-2738.

If a request for a hearing or a petition for leave to intervene is filed by the date established in the *Federal Register* notice, the Commission or the Atomic Safety and Licensing Board will rule on the request or petition, and the Commission or the designated Atomic Safety and Licensing Board will issue a notice of hearing or an appropriate order. In the event that no request for a hearing or petition is granted, the NRC may, upon completion of its evaluation and upon making the findings required under the regulations, renew the licenses without further notice.

5.1.10 What must be included in the request for a hearing or the petition to intervene?

The regulations (10 CFR 2.309) provide that a request for a hearing or a petition for leave to intervene must show the interest of the petitioner in the proceeding and how that interest may be affected by the results of the proceeding. The petition must specifically explain the reasons that intervention should be permitted, with particular reference to the following factors: 1) the nature of the petitioner's right to be made a party to the proceeding, 2) the nature and extent of the petitioner's property, financial, or other interest in the proceeding, and 3) the possible effect of any order that may be entered in the proceeding on the petitioner's interest. The petition must also identify the contentions and bases for contentions and show the hearing track under which the hearing should be conducted.

Each contention must consist of a specific statement of the issue of law or fact to be raised or controverted. The petitioner must also provide a brief explanation of the bases of each contention and a concise statement of the alleged facts or the expert opinion that supports the contention and on which the petitioner intends to rely in proving the contention at the hearing. The petitioner must also provide references to those specific sources and documents of which the petitioner is aware and on which the petitioner intends to rely to establish those facts or expert opinion. The petitioner must provide sufficient information to show that a genuine dispute exists with the applicant on a material issue of law or fact. Contentions must be limited to matters within the scope of license renewal (the action under consideration). The contention also must be one that, if proven, would entitle the petitioner to relief.

5.1.11 How do I bring safety, environmental, and security issues to the attention of the NRC?

There are two methods of reporting safety or security concerns to the NRC. The choice depends on whether the concern is considered an emergency or not.

Emergency concerns include:

- any accident involving a nuclear reactor, nuclear fuel facility, or radioactive materials,
- lost or damaged radioactive materials, and
- any threat, theft, smuggling, vandalism, or terrorist activity involving a nuclear facility or radioactive materials.

Members of the public reporting an emergency concern should call the NRC's 24-hour Headquarters Operations Center at 301-816-5100. Collect calls are accepted. All calls to this number are recorded.

Non-emergency concerns should be brought to the attention of the NRC project manager assigned to a specific plant. The list of NRC project managers is located at http://www.nrc.gov/reactors/operating/project-managers.html#pwr. This page also contains a quick link to the NRC telephone directory.

5.1.12 I might have a safety, security, or environmental issue related to a specific facility. Do I have to wait until the licensee requests license renewal to have it considered? What if the license renewal has already occurred?

Anyone who has a concern of a safety or environmental nature that applies directly to an operating facility need not wait until the licensee requests license renewal to report that concern. Concerns about a specific facility should be forwarded to the NRC project manager who is assigned to the site and listed on the NRC website at http://www.nrc.gov/reactors/operating/project-managers.html#pwr. Concerns related to license renewal should be held until (and if) the licensee submits a license renewal application. If license renewal has already been granted, then the concern should be reported to the NRC operating reactor's project manager or, if it involves an emergency as defined in the response to Question 5.1.11, to the NRC Headquarters Operation Center.

5.2 Obtaining Additional Information

5.2.1 Are documents locally available during the license renewal review?

Hard copies of documents pertinent to the environmental review are made available to the public at one or more local community libraries in the vicinity of the facility. The documents include a copy of the licensee's application containing its environmental report and a copy of the pertinent draft site-specific supplement to the generic environment impact statement on license renewal (SEIS). The location of the libraries is provided in the *Federal Register* notices related to the environmental review and can be obtained by calling the environmental project manager listed on the NRC website for each specific facility.

Documents are also available electronically through the NRC's website, as discussed in the response to Questions 5.2.3, 5.2.4, 5.2.7, and 5.2.8.

5.2.2   May I add my name to a list to receive information during the environmental review?

Members of the public may add their names to a list to receive information, including a copy of the draft and final site-specific supplement to the generic environmental impact statement on license renewal (SEIS). A sign-up sheet is available in the lobby outside the public meetings related to the environmental review. With requests for information, members of the public may also contact the NRC's environmental project manager listed on the NRC website for each specific facility.

Members of the public may add their names to a list to receive information, including a copy of the draft and final site specific supplement to the generic environmental impact statement on license renewal (SEIS).

5.2.3   Does the NRC have a website?

Yes, the NRC has a website that is updated almost daily. The website address is www.nrc.gov.

5.2.4   What kind of information on license renewal can I get from the NRC's website?

The NRC website has a special section dedicated to reactor license renewal. There is a "quick link" under the heading "Key Topics" that is shown on the right side of the NRC website's home page. This quick link takes the reader to a section dedicated to the topic of reactor license renewal. There are seven major headings or sources of information located on the Reactor License Renewal section of the website:

- Overview – a brief discussion of the license renewal process.

- License Renewal Process – a more thorough description of the application process, the environmental review, and the inspection program.

- Regulations – a description of the regulations applicable to license renewal, including the license renewal rule in 10 CFR Part 54 and the environmental regulations in 10 CFR Part 51 (links to the actual regulations are also available).

- Reactor License Renewal Guidance Documents – a description of (and links for) most of the documents discussed previously in these FAQs, including the Generic Aging Lessons Learned (GALL) report (NUREG-1801), the Standard Review Plan (SRP), and Regulatory Guides and technical reports related to license renewal.

- Public Involvement in Reactor License Renewal – a brief description of the public involvement process, including links to the schedule of upcoming public meetings, documents that are currently available for comment, and information related to any adjudication in process.

- Commission Papers Presenting Staff Recommendations – includes links to papers that the NRC staff submits to the Commissioners to inform them about matters related to license renewal (e.g., rulemaking and adjudication).

- Status of Current Applications and Industry Initiatives – includes links for each of the power plants that have submitted applications, as well as a list of anticipated future applications. For each of the power plants, there is a special section containing a link to the license renewal application and environmental report, a detailed review schedule for the specific plant, links to any completed environmental impact statements and safety evaluation reports, and a list of the NRC project managers (with their phone numbers and email addresses) for current license renewal applications in process.

5.2.5   What is the *Federal Register* and how can I get a copy of it?

The *Federal Register* is the official daily publication for rules, proposed rules, and notices of Federal agencies and organizations, as well as Executive Orders and other Presidential documents. It is published by the Office of the *Federal Register*, National Archives and Records Administration. The public can search the *Federal Register* database online at http://www.gpoaccess.gov/fr/index.html. This site contains volumes of the *Federal Register* published since 1994 (Volume 59).

> *The* Federal Register *is the official daily publication for rules, proposed rules, and notices of Federal agencies and organizations, as well as Executive Orders and other Presidential documents.*

Federal Register citations are commonly given in a form that states the volume first and then, after the acronym *FR*, the page number: e.g., 60 FR 22461, indicating that it is volume 60 and page 22461. Searches on the Government Printing Office (GPO) access site can be conducted by *Federal Register* date, volume, and page or by key word (http://www.gpoaccess.gov/index.html). Other options for obtaining the *Federal Register* include purchasing a subscription (instructions are on the GPO website) or obtaining issues from a local Federal depository library. Locations of such libraries are also given on the website.

Copies of *Federal Register* notices that deal with license renewal are also located on the NRC website: http://www.nrc.gov/reactors/operating/licensing/renewal/applications.html. They are listed by the name of each facility that has applied for a renewed license.

5.2.6   How can I get a copy of the *Code of Federal Regulations* dealing with license renewal?

The license renewal regulations in Title 10, Energy, in the *Code of Federal Regulations* can be viewed and printed from the NRC website at http://www.nrc.gov/reading-rm/doc-collections/cfr/. In addition, copies of the *Code of Federal Regulations* may be purchased from the Superintendent of Documents at the U.S. Government Printing Office or the National Technical Information Service in Springfield, VA. The

Federal Register

contact information for these two locations is given in the response to Question 5.2.9.

5.2.7 How does a member of the public obtain a copy of a license renewal application for a specific nuclear power plant?

The *Federal Register* notice that indicates that the NRC has received an application from a specific site also provides information on how the public can access the application. Copies of the application are available electronically on the NRC web site at http://www.nrc.gov/reactors/operating/licensing/renewal/applications.html. The application is also available electronically from the NRC's Agency-wide Documents Access and Management System (ADAMS). The ADAMS Public Electronic Reading Room is accessible from the NRC website at http://www.nrc.gov/NRC/ADAMS/index.html. In addition, a copy or copies are available to local residents at one or two local libraries in the vicinity of the facility. The local library at which copies are available are identified in the *Federal Register* notice related to the environmental review, as discussed in the response to Question 5.2.1.

5.2.8 How do I get a copy of a draft site-specific supplement to the generic environmental impact statement on license renewal (SEIS) related to a particular facility?

A single copy of each NRC draft SEIS is free, to the extent of availability, upon written request to the following address:

Office of the Chief Information Officer,
Reproduction and Distribution Services Section
U.S. Nuclear Regulatory Commission
Washington, DC 20555-0001
E-mail: DISTRIBUTION@nrc.gov
Facsimile: 301-415-2289

Members of the public who sign up at the public scoping meeting to receive a copy of the draft SEIS for that specific facility will automatically be sent a copy once the draft SEIS is published. A copy is also available to local residents at the local libraries identified in the *Federal Register* notice, as discussed in the response to Question 5.2.1.

In addition, the draft SEIS is available for review from the NRC website at the following location: http://www.nrc.gov/reading-rm/doc-collections/nuregs/staff/sr1437.

5.2.9 How can I get a copy of the *Generic Environmental Impact Statement for License Renewal of Nuclear Plants*, NUREG-1437 (GEIS), and a final specific supplement related to a particular facility?

Copies of the GEIS and the site-specific supplement to the generic environmental impact statement on license renewal (SEIS) can be obtained from NRC's website at the following location: http://www.nrc.gov/reading-rm/doc-collections/nuregs/staff/sr1437/.

Copies can also be purchased from either of the following sources:

The Superintendent of Documents
U.S. Government Printing Office
Mail Stop SSOP
Washington, DC 20402-0001

Internet: bookstore.gpo.gov
Telephone: 202-512-1800
Fax: 202-512-2250

or

The National Technical Information Service
Springfield, VA 22161-0002

www.ntis.gov
1-800-553-6847 or, locally, 703-605-6000

5.2.10   How can I get answers to additional questions that were not addressed in this document?

Members of the public are invited to plant-specific public meetings, where NRC staff members are available to answer both generic and site-specific questions (see also the response to Question 5.1.4). In addition, many answers to questions that are not included in this document can be found on the NRC website, www.nrc.gov. The NRC has developed a number of frequently asked question documents, as well as informational brochures and fact-sheets. For plant-specific safety and environmental questions related to a license renewal application, members of the public can contact the safety and/or environmental project manager assigned by the NRC for the license renewal review for the specific plant. The name, phone number, and email address for each of the NRC safety and environmental project managers is given on the NRC website, as well as in *Federal Register* notices and at the public meetings. The NRC safety and environmental project managers can either answer questions or direct callers to the appropriate person at the NRC.

6.0 Bibliography of Published Material Relevant to License Renewal

10 CFR Part 51. "Environmental Protection Regulations for Domestic Licensing and Related Regulatory Functions."

10 CFR Part 54. "Requirements for Renewal of Operating Licenses for Nuclear Power Plants."

40 CFR Part 1500. "NEPA and Agency Planning."

69 FR 52040. "Policy Statement on the Treatment of Environmental Justice Matters in NRC Regulatory and Licensing Actions."

Atomic Energy Act of 1954, as amended, 16 USC 1531, et seq.

American Cancer Society (ACS). 2004. Cancer Facts and Figures – 2004. http://www.cancer.org/downloads/STT/CAFF_finalPWSecured.pdf

Connecticut Academy of Science and Engineering, Spring 2001. Bulletin of the Connecticut Academy of Science and Engineering, Volume 16.2. Hartford, Connecticut.

Bureau of Environmental Epidemiology, Florida Department of Health. 2001. Report Concerning Cancer Rates in Southeastern Florida. In Letter from David R. Johnson to Interested Parties. July 18, 2001. http://www.fpl.com/about/nuclear/contentsfdh_report.shtml

Illinois Public Department of Health, Fall 2000. Health and Hazardous Substances Registry Newsletter. Illinois Public Health, Division of Epidemiologic Studies, Illinois Department of Public Health, 605 W. Jefferson St., Springfield, Illinois.

International Commission on Radiological Protection (ICRP). 1991. *1990 Recommendations of the International Commission on Radiological Protection*. ICRP Publication No. 60. Annals of the ICRP. Elsevier.

National Cancer Institute (NCI). 1990. *Cancer in Populations Living Near Nuclear Facilities*. Bethesda, Maryland.

National Research Council. 1990. *Health Effects of Exposure to Low Levels of Ionizing Radiation. BEIR V.* Committee on the Biological Effects of Ionizing Radiations. Board on Radiation Effects Research Commission of Life Sciences. National Research Council. National Academy Press, Washington, D.C.

National Research Council. 2006. *Health Risks from Exposure to Low Levels of Ionizing Radiation: BEIR VII Phase II*. Committee to Assess Health Risks from Exposure to Low Levels of Ionizing Radiation. National Research Council. National Academy Press, Washington, D.C.

National Council on Radiation Protection and Measurement (NCRP). 1987. *Ionizing Radiation Exposures of the Population of the United States*. Report No. 93. Washington, D.C.

National Council on Radiation Protection and Measurements (NCRP). 1988. *Public Radiation Exposure from Nuclear Power Generation in the United States.* Report No. 92. Washington, D.C.

National Council on Radiation Protection and Measurement (NCRP). 1988. *Exposure of the Population in the United States and Canada from Natural Background Radiation.* Report No. 94. Washington, D.C.

Talbott, Evelyn O, A.O. Youk, K.P. McHugh-Pemu, and J.V. Zborowski. 2003. Long Term Follow-up of the Residents of the Three Mile Island Accident Area: 1979-1998. In Environmental Health Perspectives, Volume 111, Number 3. March 2003. Pages 341-348.

U.S. Nuclear Regulatory Commission (NRC). 1996. *Generic Environmental Impact Statement for License Renewal of Nuclear Plants(GEIS).* NUREG-1437. Washington, D.C.

U.S. Nuclear Regulatory Commission (NRC). 1999. *Standard Review Plans for Environmental Reviews for Nuclear Power Plants – Supplement 1: Operating License Renewal.* NUREG-1555, Supplement 1. Washington, D.C.

U.S. Nuclear Regulatory Commission (NRC). 2000. *Preparation of Supplemental Environmental Reports for Applications to Renew Nuclear Power Plant Operating Licenses.* Supplement 1 to Regulatory Guide 4.2. Washington, D.C.

U.S. Nuclear Regulatory Commission (NRC). 2001. *Standard Review Plan for Review of License Renewal Applications for Nuclear Power Plants.* NUREG-1800. Washington, D.C.

U.S. Nuclear Regulatory Commission (NRC). 2001. *Generic Aging Lessons Learned (GALL) Report.* NUREG-1801. Washington, D.C.

U.S. Nuclear Regulatory Commission (NRC). 2001. *Standard Format and Content for Applications to Renew Nuclear Power Plant Operating Licenses.* Regulatory Guide 1.188. Washington, D.C.

U.S. Nuclear Regulatory Commission (NRC). 2005. *License Renewal Inspections.* Inspection Manual 71002. Washington, D.C.

U.S. Nuclear Regulatory Commission (NRC). 2005. *Policy and Guidance for Licence Renewal Inspection Programs.* MC-2516. Washington, D.C.

U.S. Nuclear Regulatory Commission (NRC). Office of Nuclear Reactor Regulation Office *Procedural Guidance for Preparing Environmental Assessments and Considering Environmental Issues.* Instructor LIC-203. Washington, D.C.

Appendix A

Environmental Issues for License Renewal
of Nuclear Power Plants

Appendix A
Environmental Issues for License Renewal of Nuclear Power Plants
(10 CFR Part 51, Subpart A, Appendix B)

The impact evaluation performed by the staff and presented in the *Generic Environmental Impact Statement for License Renewal of Nuclear Plants*, NUREG-1437 (GEIS), identified 92 environmental issues that needed to be considered for the license renewal evaluation for power reactors in the U.S. These issues are numbered consecutively in the following table (Table A-1) along with short descriptions of the issue and the category type. For each of the identified 92 issues, the staff evaluated existing data relative to all operating power plants throughout the U.S. From this evaluation, the staff determined which issues could be considered generically and which issues do not lend themselves to a generic classification. The GEIS divided the 92 issues that were assessed into two categories: one for generic issues (which are termed "Category 1 issues") and the other for site-specific issues (termed "Category 2").

Category 1 (generic) issues are those that meet all of the following criteria:

1) The environmental impacts associated with the issue have been determined to apply either to all plants or, for some issues, to plants having a specific type of cooling system or other specified plant or site characteristic.

2) A single significance level (i.e., SMALL, MODERATE, or LARGE) has been assigned to the impacts (except for collective offsite radiological impacts from the fuel cycle and from high-level waste and spent fuel disposal). (See response to Question 4.2.4.13 for a definition of the levels of significance.)

3) Mitigation of adverse impacts associated with the issue has been considered in the analyses, and it has been determined that additional plant-specific mitigation measures are not likely to be sufficiently beneficial to warrant implementation.

Category 1 issues are termed "generic" because the conclusions related to their environmental impacts were found to be common to all plants (or, in some cases, to plants having specific characteristics such as a specific type of cooling system). In such cases, a single level of significance can be assigned to them, and mitigation is not likely to be beneficial. Issues that are found to be "generic" are not reevaluated in the site-specific supplement to the generic environmental impact statement on license renewal (SEIS) because the conclusions reached would be the same as in the GEIS, unless new and significant information is found that would lead the NRC staff to reevaluate the GEIS's conclusions.

Category 2 issues are those in which the GEIS offers no generic conclusion. These issues require a site-specific review. For each of the Category 2 issues, the staff evaluates site-specific data provided by the licensee, other Federal agencies, state agencies, and local government agencies as well as information from the open literature and from members of the public. From all these data, the staff makes a site-specific evaluation of the particular issues and presents its analyses and conclusions in the SEIS for the facility.

The GEIS evaluates 92 environmental issues, and, of these, 69 were found to be generic (Category 1) and 23 issues needed a site-specific review and analysis, with 21 of these considered to be Category 2 issues. The remaining 2 issues (environmental justice and chronic effects of electromagnetic fields) were not categorized and are addressed in the site-specific analysis.

Table A-1. NEPA Issues for License Renewal of Nuclear Power Plants

Issue Number	Issue Title	Category
1	Impacts of refurbishment on surface-water quality	1
2	Impacts of refurbishment on surface-water use	1
3	Altered current patterns at intake and discharge structures	1
4	Altered salinity gradients	1
5	Altered thermal stratification of lakes	1
6	Temperature effects on sediment transport capacity	1
7	Scouring caused by discharged cooling water	1
8	Eutrophication	1
9	Discharge of chlorine or other biocides	1
10	Discharge of sanitary wastes and minor chemical spills	1
11	Discharge of other metals in wastewater	1
12	Water use conflicts (plants with once-through cooling systems)	1
13	Water use conflicts (plants with cooling ponds or cooling towers using make-up water from a small river with low flow)	2
14	Refurbishment	1
15	Accumulation of contaminants in sediments or biota	1
16	Entrainment of phytoplankton and zooplankton	1
17	Cold shock	1
18	Thermal plume barrier to migrating fish	1
19	Distribution of aquatic organisms	1
20	Premature emergence of aquatic insects	1
21	Gas supersaturation (gas bubble disease)	1
22	Low dissolved oxygen in the discharge	1
23	Losses from predation, parasitism, and disease among organisms exposed to sublethal stresses	1
24	Stimulation of nuisance organisms	1
25	Entrainment of fish and shellfish in early life stages	2
26	Impingement of fish and shellfish	2
27	Heat shock	2
28	Entrainment of fish and shellfish in early life stages	1
29	Impingement of fish and shellfish	1
30	Heat shock	1
31	Impacts of refurbishment on groundwater use and quality	1
32	Groundwater-use conflicts (potable and service water; plants that use <100 gpm).	1
33	Groundwater-use conflicts (potable and service water, and dewatering; plants that use > 100 gpm)	2
34	Groundwater-use conflicts (plants using cooling towers withdrawing makeup water from a small river)	2
35	Groundwater-use conflicts (Ranney wells)	2
36	Groundwater quality degradation (Ranney wells)	1
37	Groundwater quality degradation (saltwater intrusion)	1
38	Groundwater quality degradation (cooling ponds in salt marshes)	1
39	Groundwater quality degradation (cooling ponds at inland sites)	2
40	Refurbishment impacts	2
41	Cooling tower impacts on crops and ornamental vegetation	1
42	Cooling tower impacts on native vegetation	1
43	Bird collisions with cooling towers	1
44	Cooling pond impacts on terrestrial resources	1
45	Power line right-of-way management (cutting and herbicide application)	1
46	Bird collisions with power lines	1
47	Impacts of electromagnetic fields on flora and fauna (plants, agricultural crops, honeybees, wildlife, livestock)	1

Table A-1. (contd)

Issue Number	Issue Title	Category
48	Flood plains and wetland on power line right-of-way	1
49	Threatened or endangered species	2
50	Air quality during refurbishment (nonattainment and maintenance areas)	2
51	Air-quality effects of transmission lines	1
52	Onsite land use	1
53	Power line right-of-way	1
54	Radiation exposures to the public during refurbishment	1
55	Occupational radiation exposures during refurbishment	1
56	Microbial organisms (occupational health)	1
57	Microbiological organisms (public health) (plants using lakes or canals or cooling towers that discharge into a small river)	2
58	Noise	1
59	Electromagnetic fields, acute effects (electric shock)	2
60	Electromagnetic fields, chronic effects	N/A
61	Radiation exposures to public (license renewal term)	1
62	Occupational radiation exposures (license renewal term)	1
63	Housing impacts	2
64	Public services: public safety, social services, and tourism and recreation	1
65	Public services: public utilities-water supply	2
66	Public services: education (refurbishment)	2
67	Public services: education (license renewal term)	1
68	Offsite land use (refurbishment)	2
69	Offsite land use (license renewal term)	2
70	Public Services, transportation	2
71	Historic and archaeological resources	2
72	Aesthetic impacts (refurbishment)	1
73	Aesthetic impacts (license renewal term)	1
74	Aesthetic impacts of transmission lines (license renewal term)	1
75	Design-basis accidents (DBAs)	1
76	Severe Accidents	2
77	Offsite radiological impacts (individual effects from other than the disposal of spent fuel and HLW)	1
78	Offsite radiological impacts (collective effects)	1
79	Offsite radiological impacts (spent fuel and HLW)	1
80	Nonradiological impacts of the uranium fuel cycle	1
81	Low-level waste storage and disposal	1
82	Mixed waste storage and disposal	1
83	Onsite spent fuel	1
84	Nonradiological waste	1
85	Transportation	1
86	Radiation Doses	1
87	Waste Management	1
88	Air Quality	1
89	Water Quality	1
90	Ecological Resources	1
91	Socioeconomic Impacts	1
92	Environmental Justice	N/A

NRC FORM 335 (9-2004) NRCMD 3.7	U.S. NUCLEAR REGULATORY COMMISSION	1. REPORT NUMBER (Assigned by NRC, Add Vol., Supp., Rev., and Addendum Numbers, if any.)
BIBLIOGRAPHIC DATA SHEET *(See instructions on the reverse)*		NUREG-1850

2. TITLE AND SUBTITLE		3. DATE REPORT PUBLISHED	
Frequently Asked Questions on License Renewal of Nuclear Power Reactors		MONTH	YEAR
		March	2006
		4. FIN OR GRANT NUMBER	

5. AUTHOR(S)	6. TYPE OF REPORT
	Final
	7. PERIOD COVERED *(Inclusive Dates)*

8. PERFORMING ORGANIZATION - NAME AND ADDRESS *(If NRC, provide Division, Office or Region, U.S. Nuclear Regulatory Commission, and mailing address; if contractor, provide name and mailing address.)*

Division of License Renewal
Office of Nuclear Reactor Regulation
U.S. Nuclear Regulatory Commission
Washington, DC 20555-001

9. SPONSORING ORGANIZATION - NAME AND ADDRESS *(If NRC, type "Same as above"; if contractor, provide NRC Division, Office or Region, U.S. Nuclear Regulatory Commission, and mailing address.)*

Same as above.

10. SUPPLEMENTARY NOTES

11. ABSTRACT *(200 words or less)*

This report, through a question-and-answer format, provides staff responses to frequently asked questions on the license renewal process for commercial, nuclear power reactors. The questions were taken from a variety of sources over the past several years, including written inquiries to the NRC and questions asked at public meetings and during informal discussions with the NRC staff. The NRC staff attempted to provide answers in a clear and non-technical form.

This document contains a definition of license renewal including information related to the timing and scheduling of the license renewal process. It discusses the NRC's role in reviewing, approving, or denying license renewal and the regulatory basis for the review. Because the public usually encounters the license renewal process in conjunction with the environmental review, this document is primarily focused on the environmental review process and on related issues such as alternatives to license renewal and human health issues. However, other aspects of license renewal are addressed, including questions related to the safety reviews, the storage and disposal of spent nuclear fuel, and security issues. There are also responses to questions related to public involvement and to finding sources of additional information on the license renewal process.

12. KEY WORDS/DESCRIPTORS *(List words or phrases that will assist researchers in locating the report.)*	13. AVAILABILITY STATEMENT
FAQ, Frequently asked questions, License, Renewal, License Renewal, Environmental, ER, Environmental review, Review, NEPA, Impact assessment, Public involvement, EA, Environmental assessment, EIS, Environmental Impact Statement, Process	unlimited
	14. SECURITY CLASSIFICATION
	(This Page)
	unclassified
	(This Report)
	unclassified
	15. NUMBER OF PAGES
	16. PRICE

www.ingramcontent.com/pod-product-compliance
Lightning Source LLC
Chambersburg PA
CBHW081501170526
45166CB00008B/2514